D1064592

Transport Processes in Lakes and Oceans

MARINE SCIENCE

Coordinating Editor: Ronald J. Gibbs, *University of Delaware*

Volume 1 – *Physics of Sound in Marine Sediments*
Edited by Loyd Hampton

An Office of Naval Research symposium
Consulting Editors: Alexander Malahoff and Donald Heinrichs
Department of the Navy

Volume 2 – *Deep-Sea Sediments: Physical and Mechanical Properties*
Edited by Anton L. Inderbitzen

An Office of Naval Research symposium
Consulting Editors: Alexander Malahoff and Donald Heinrichs
Department of the Navy

Volume 3 – *Natural Gases in Marine Sediments*
Edited by Isaac R. Kaplan

An Office of Naval Research symposium
Consulting Editors: Alexander Malahoff and Donald Heinrichs
Department of the Navy

Volume 4 – *Suspended Solids in Water*
Edited by Ronald J. Gibbs

An Office of Naval Research symposium
Consulting Editors: Alexander Malahoff and Donald Heinrichs
Department of the Navy

Volume 5 – *Oceanic Sound Scattering Prediction*
Edited by Neil R. Andersen and Bernard J. Zahuranec

An Office of Naval Research symposium

Volume 6 – *The Fate of Fossil Fuel CO_2 in the Oceans*
Edited by Neil R. Andersen and Alexander Malahoff

An Office of Naval Research symposium

Volume 7 – *Transport Processes in Lakes and Oceans*
Edited by Ronald J. Gibbs

Consulting Editor: Richard P. Shaw

A Continuation Order Plan is available for this series. A continuation order will bring delivery of each new volume immediately upon publication. Volumes are billed only upon actual shipment. For further information please contact the publisher.

Transport Processes in Lakes and Oceans

Edited by
Ronald J. Gibbs
Marine Studies Center
University of Delaware

Consulting Editor
Richard P. Shaw
Department of Engineering Sciences
State University of New York at Buffalo

PLENUM PRESS • NEW YORK AND LONDON

Library of Congress Cataloging in Publication Data

Symposium on Transport Processes in the Oceans, Atlantic City, 1976.

Transport processes in lakes and oceans.
(Marine Science; v. 7)
Proceedings of a symposium held at the 82nd national meeting of the American Institute of Chemical Engineers, and other selected papers.
1. Ocean circulation—Congresses. 2. Water chemistry—Congresses. 3. Ocean atmosphere interaction—Congresses. 4. Marine pollution. I. Gibbs, Ronald J., 1933-
II. American Institute of Chemical Engineers. III. Title.
GC228.5.594 1976 551.4'701 77-16279
ISBN 0-306-35507-8

Proceedings of a Symposium on Transport Processes in the Oceans held at the Eighty-Second National Meeting of the American Institute of Chemical Engineers, Atlantic City, New Jersey, August 29—September 1, 1976, and other selected papers

A Division of Plenum Publishing Corporation
227 West 17th Street, New York, N.Y. 10011

Printed in the United States of America

PREFACE

The Eighty-Second National Meeting of the American Institute of Chemical Engineers, held in Atlantic City, New Jersey, from August 29 through September 1, 1976, had as one of its themes the topic of transport processes. One of the sessions related to this theme was "Transport Processes in the Oceans" chaired by R. P. Shaw and R. J. Gibbs. This session was devoted to the study of transport processes and their hydrodynamic modeling in large water bodies such as oceans and lakes; transport of both dissolved and solid material was considered. The interest developed at the session led to the conclusion that the papers presented there should be published as a set rather than dispersed among the various technical journals that represent the wide variety of technical affiliations of the authors. This variety, in fact, is typical of this particular field with contributors identified as chemical engineers, civil engineers, environmental engineers, mechanical engineers, oceanographers and applied mechanicians to name just a few. Such an interdisciplinary area requires more effort in keeping abreast of developments than do the traditional areas, since new material may be developed and presented in a wide range of technical journals and professional meetings.

During the development of this volume, it seemed clear that transfer processes, i.e., the means by which material enters and leaves the hydro-environment, were of interest equal to that of transport processes, i.e., how material moves about after it is in the hydro-environment. To add this dimension of interest to the volume, several papers arising from a seminar series held at the State University of New York at Buffalo during 1973, entitled "Interfacial Transfer Processes in Water Resources," were added. It is felt that this provides the optimum combination of background material and current research results in the areas of transfer and transport processes in large water bodies.

Richard P. Shaw

CONTENTS

TRANSPORT PROCESSES IN LAKES AND OCEANS

Ronald J. Gibbs

College of Marine Studies

University of Delaware

INTRODUCTION

The researchers studying transport processes in lakes and oceans include physical, chemical and biological oceanographers, geologists, meteorologists, environmentalists and civil, chemical and mechanical engineers. These comprise an exceedingly varied group of disciplines, having the distinct advantage that problems can be approached from a wide range of viewpoints. However, it can also lead to serious disadvantages, with probably the most serious and wasteful being that, due to a number of factors, a researcher might not take advantage of previous work accomplished in a disciplinary field other than that of his own interest. Some of these factors are, of course, the different training researchers may have had and the approaches to research in the various disciplines. Equally important is the fact that the findings of each discipline are usually published in the specialty journals of the particular discipline and, in general, may not be read by workers in another discipline. Findings from research may likewise, not be presented in terms readily understood by researchers in another discipline. A discussion of the broad topic of transport processes in lakes and oceans can be subdivided into three main areas:

I. Water Transportation

II. Transportation of Dissolved Material

A. Conservative materials

B. Non-conservative materials

III. Transportation of Solid Material

A. Inorganic material

B. Organic material

WATER TRANSPORTATION

In the study of transport processes, the subject of water circulation in the deep sea has received greatest attention from research in the physical oceanographic community. The reason for emphasis on deep-sea water circulation is twofold. First, this area was thought not to have the complex boundary conditions encountered in studies of coastal and estuarine circulation. Secondly, over the past several decades, funding for deep-sea work has been more readily available than has funding for coastal studies. (For example, the Office of Naval Research's physical oceanography program funded mainly deep-water research.) As a consequence of this emphasis, there remains in the coastal region a wealth of topics to be investigated. A whole spectrum of problems requiring investigation develops as waves, tides, and currents interact with the bottom in the continental shelf and slope regions, as well as with the coast line itself.

In the past, the research accomplished regarding water circulation on the continental shelf regions was mainly a byproduct of geologists studying sediment transport or engineers studying erosion and siltation. In general, there was no concerted effort to understand in detail the overall circulation of the shelf regions. A recent text that may aid in understanding water circulation in estuaries and coastal regions is that by Officer (1976). A volume covering general physical oceanography is Neuman and Pierson (1966). In recent years, more funding has been directed to this region by the National Science Foundation for such projects as research on coastal upwelling. Funding from the Bureau of Land Management has supported environmental baseline studies along the coastal areas.

The wide variety of research topics in the area of water transportation available for investigation is evident in such journals as *Deep-Sea Research*, *Journal of Marine Research*, *Journal of Geophysical Research*, *Journal of Physical Oceanography* for shelf and deep-ocean research and in various geological and engineering journals for research on coastal interaction. A recent symposium assessing research on geophysics of estuaries sponsored by the National Academy of Sciences (Officer, 1977) should also be helpful. Included in the symposium volume are chapters dealing with various aspects of circulation in estuaries and coastal regions reviewing research accomplished in a particular aspect and setting forth proposals for future research.

TRANSPORTATION OF DISSOLVED MATERIAL

Those dissolved substances that are generally non-reactive, or conservative, upon entering the ocean can, in transport models, be considered as traveling with their water mass. The concentration of these substances when mixed will be controlled by the simple dilution between the two end members considered (Figure 1), giving a straight-line dilution relationship.

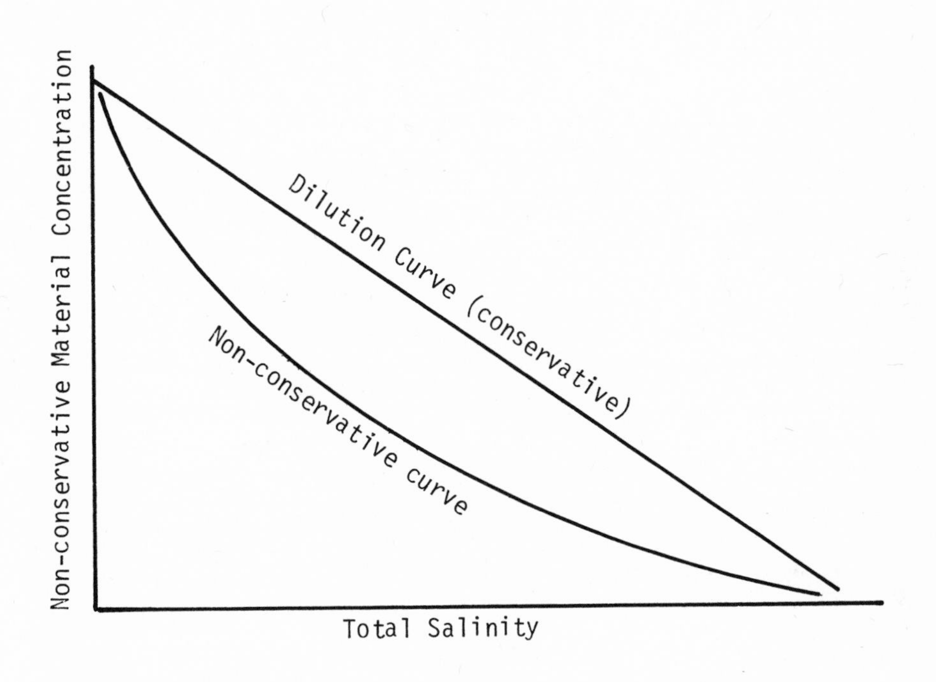

Figure 1. The relationship of total salinity and conservative and non-conservative material concentrations.

Complications arise with the materials that are associated with two natural mechanisms that cause them to change concen tration with time by a mechanism other than dilution. One of these mechanisms is the conversion of this material from the dissolved phase into solid biological material by biological creatures. In general, this mainly affects the nutrients: the compounds of N, P and Si, along with O_2 and CO_2 and a number of minor nutrients and natural waste products. The case of Si has been studied extensively in estuaries and river/ocean interfaces (Liss and Pointon, 1973; Fanning and Pilson, 1973). The general conclusion is that Si is utilized by the biota and is preferentially removed from solution. This results in the concentration of Si verses total salinity exhibiting a curve that decreases faster than that exhibited by a generally conservative substance, for example, when Si-rich river water is mixed with Si-poor sea water (Figure 1). The same non-conservative relationship occurs with most of the other nutrients.

A second mechanism causing non-conservative behavior is the inorganic precipitation process related to a changing physiochemical environment. One example of this process is the inorganic precipitation of dissolved organic material and dissolved iron when a river flows into the ocean (Beck et al., 1974; Sholkovitz, 1976).

The impact of man on non-conservative transport complicates an already complex model. Not only does man add nutrients to the aqueous environment (sometimes in such massive quantities that the system is thrown seriously out of balance to the positive side) but he also adds biologically toxic materials, resulting in a negative effect by killing the creatures that caused the previously non-conservative reaction. In many instances man actually adds both positive-effect nutrients and negative-effect toxic materials to the natural system simultaneously, producing a transport system exceedingly complex to understand and model.

The state of this area of research has not progressed very far because of the complexities to be considered. The literature covering work accomplished is scattered among a dozen journals in a number of fields. The proceedings of the International Estuarine Research Federation Symposia (Wiley, 1976a and b) provide a good starting point for interested investigators inasmuch as almost all of the relevant subject areas are covered and an abundance of references is given.

TRANSPORTATION OF SOLID MATERIAL

The problem of solid-material transportation can be examined by considering the organic and the inorganic material.

In general, the inorganic solid material does not travel with the water mass. The particles of this material can be dissolved in the ocean; for example, calcium carbonate particles settling into the deeper cold waters of the ocean (Peterson, 1966; Berger and Piper, 1972). The particles of inorganic solid material can flocculate upon entering sea water, resulting in larger particles which, in turn, settle faster (Whitehouse et al., 1960; Edzwald and O'Melia, 1975; Gibbs, 1977). In addition, particles of inorganic solid material can settle from the water column at varying rates related to particle size; they can be removed from the water column by ingestion by the biota followed by defecation with the resulting large particles having higher settling velocity; or, they can by added to the water column by being stirred up from the bottom. An understanding of these processes can be expanded with study of the numerous references in Swift et al. (1972), Meade (1972), Gibbs (1974), Stanley and Swift (1976), and Officer (1977).

A complicating phenomenon in the natural system arises from the *in situ* production of solid organic particles by biological processes, both as living creatures and as the remains of creatures. The rate of destruction for the soft parts of these creatures and their skeletal materials varies. The biological processes of production and destruction, which are consistently present, are superimposed on the inorganic processes. The total quantity of solid material produced by biological processes varies seasonally and by geographical area; the processes themselves are controlled by water temperature, nutrient levels, and amount of light.

Transport of solid material in lakes and oceans is seen, therefore, to be a very complex system. Progress in the study of transportation of the organic solid material is as lacking as in that of the inorganic material, with most of the same processes in need of investigation for both types of materials. The literature on research accomplished in transportation of solid material is distributed among journals too numerous to mention. However, there are several volumes that consider nearly all areas of this topic in sufficient detail to indicate what has been accomplished and what aspects should be investigated (Gibbs, 1974; Officer, 1977; Wiley, 1976a and b). These books also contain many references that will aid researchers contemplating work in this area.

CONCLUSION

While the subject of transport processes in lakes and oceans has been considered here with regard to water transportation, transportation of dissolved material and transportation of solid material, there are, in reality, no boundaries among the three areas. Obviously, the processes of water transport must be understood in order to be able to attack the more complex problems of transportation of dissolved and solid materials. The opportunities for research in these complex areas is unlimited, with numerous significant problems available for investigation. The funding situation is improving; however, this is a perplexing and difficult subject area in which to make significant progress successfully.

REFERENCES

Beck, K. C., J. H. Reuter, and E. M. Perdue, 1974, Organic and inorganic geochemistry of some coastal plain rivers of the southeastern United States: Geochim. et Cosmochim. Acta, v. 38, p. 341-364.

Berger, W. H. and D. J. W. Piper, 1972, Planktonic foraminifera: different settling, dissolution and redeposition: Limnology and Oceanology, v. 17, p. 275-287.

Edzwald, J. K. and C. R. O'Melia, 1975, Clay distribution in recent estuarine sediments: Clays and Clay Minerals, v. 23, p. 39-44.

Fanning, K. A. and M. E. Q. Pilson, 1973, The lack of inorganic removal of dissolved silica during river-ocean mixing: Geochim. et Cosmochim. Acta, v. 37, p. 2405-2415.

Gibbs, R. J. (ed.), 1974, Suspended Solids in Water. New York, Plenum Press, 320 pp.

Gibbs, R. J., 1977, Clay mineral segregation in the marine environment: J. Sed. Pet., v. 47, p. 237-243.

Liss, P. S. and M. J. Pointon, 1973, Removal of dissolved boron and silicon during estuarine mixing of sea and river waters: Geochim. et Cosmochim. Acta, v. 37, p. 1493-1498.

Meade, R. H., 1972, Transport and deposition of sediments in estuaries: in Environmental framework of coastal plain estuaries. B. W. Nelson (ed.), Geol. Soc. Amer. Mem. 133, p. 91-117.

Neuman, G. and W. J. Pierson, 1966, Principles of physical oceanography, Prentice-Hall, Inc., N. Y. 545 pp.

Officer, C. B., 1976, Physical Oceanography of Estuaries, John Wiley and Sons, 465 pp.

Officer, C. B. (ed.), 1977, Studies in Geophysics: National Assessment of Geophysics of Estuaries. Symposium held in conjunction with the National Meeting of the American Association for the Advancement of Science, Boston, February 18-24, 1976. National Academy of Sciences, in press.

Peterson, M. N. A., 1966, Calcite: rates of dissolution in a vertical profile in the central Pacific: Science, v. 154, p. 1542-1544.

Sholkovitz, E. R., 1976, Flocculation of dissolved organic and inorganic matter during the mixing of river water and seawater: Geochim. et Cosmochim Acta, v. 40, p. 831-845.

Stanley, J. D. and D. J. P. Swift, 1976, Marine Sediment Transport and Environmental Management, John Wiley and Sons, 602 pp.

Swift, D. J. P., D. B. Duane and O. H. Pilkey, 1972, Shelf Sediment Transport Processes and Patterns. Dowden, Hutchinson, and Ross, Inc., 656 pp.

Whitehouse, G., L. M. Jeffrey, and J. D. Debbrecht, 1960, Differential settling tendencies of clay minerals in saline waters: Clays and Clay Minerals, v. 7, p. 1-79.

Wiley, M. (ed.), 1976a, Estuarine processes, Vol. I, Uses, stresses and adaptation to the estuary, Academic Press, 583 pp.

Wiley, M. (ed.), 1976b, Estuarine processes, Vol. II, Circulation, sediments and transfer of material in the estuary, Academic Press, 428 pp.

SOLID-SOLUTION INTERFACE:

ITS ROLE IN REGULATING THE CHEMICAL COMPOSITION OF NATURAL WATERS

C. P. Huang

University of Delaware

ABSTRACT

The presence of suspended particles in natural waters can significantly affect the fate of most chemical constituents. Selective adsorption of chemical species onto suspended particles followed by differential coagulation of the suspended particulate signifies the importance of solid-solution interface in regulating the chemical composition of natural waters. The above interfacial reactions are governed by the intrinsic properties of the suspended particles, the nature of the solutes and the physical-chemical characteristics of the aquatic environment.

INTRODUCTION

The presence of suspended particles in the aquatic systems has a significant influence over the dynamic distribution of the particles per sec and the chemical composition of the whole system.

The production and distribution of the primary biological particles, for instance, have direct affect on the population dynamics of the bio-particles of higher order, and fishes [Sheldon et al., 1973; Riley, 1963]. Moreover, the presence of phytoplankton also has significant control ober the concentration of specific micronutrients, e.g. nitrogen, phosphate, calcium, and silicate, in the natural bodies of water.

It is generally known that solutes tend to cumulate at the solid-solution interface. It is evident from many reported field and laboratory data that the concentration of chemical species, such as trace metals, hydrocarbons, phosphate, silicate, in the bottom sediments or in the particulate solids of aquatic system is at least one or more orders of magnitude greater than that found in the solution part [Stumm and Leckie, 1971; Leckie and James, 1974]. Chemical association at solid-solution interface is brought by specific chemical bonds, by electrostatic attraction or by bridging formation. Interfacial reactions tend to destabilize the suspended particles and thereby result in material transport into the aquatic sediments [Morgan and Stumm, 1964].

The major objective of this present work is to describe the fundamental aspects of solid-solution interface and its control over the aqueous chemical constituents such as trace metals, inorganic anions, and hydrocarbons in the aquatic system. In order to appreciate fully the significance of solid-solution interface, a description of the sources, origins, and nature of suspended particles in the aquatic environment is also discussed.

AQUEOUS SUSPENDED PARTICLES

The Origins and Sources of Suspended Particles

Aqueous suspended solids can originate from various sources and are brought into natural waters by a number of mechanisms. Basically, particulates are derived from either natural (e.g., weathering, volcanic enamation, hydrothermal activities, biosynthesis) or man-influenced activities (e.g., waste discharge, dredging, agricultural and mining operations). Generally, particles are brought into aquatic systems from outer sources or in situ formation by physical, chemical, and biological mechanisms.

Natural Sources. Natural weathering forces cause rock deterioration and contribute a great portion of suspended particles of various kinds, e.g. oxides, hydroxides, carbonates, clays, heavy minerals, to all surface waters [Turekian, 1968]. Most of the weathering products have a grain-size smaller than 20 μ diameter [Mason, 1966]. The extent and dynamics of weathering are influenced by the prevailing physical conditions in the geographical area of interest.

Materials from volcanic emanation are usually of the volatile form. But, in addition, gaseous emanation from magma collect, transport, and deposit many other elements (e.g., magnetite, hematite, molybdemite, pyrite, realgar, galena, sphalerite, covellite, sal ammoniae, ferric chloride, and many others) [Mason, 1966]. The congruent and incongruent solubility characteristics of minerals play a significant part in the in situ formation of suspended particles. Under favorable environmental conditions (e.g., temperature, pH, reducing-oxidation level, and concentrations of solute) specific solids are found in the aquatic environment [Krumbein and Garrels, 1952].

Aside from the solid particles, Riley [1963, 1970] observed that adsorption of organic substances onto gas bubbles can also create a stable particulate whole size ranging from 25 to 50 μ in the oceanic water. A similar mechanism is also found operative in the fresh water. Dissolved organic matters derived from decayed leaves are precipitated by turbulence and cations, e.g. Ca^{+2}, to form organic particulates. The rate of precipitation and the size of particles (usually ranging from 5 to 35 μ) depend on leaf species and water chemistry [Lush and Hynes, 1973].

Liquid droplet is another type of suspended particle found in the aqueous environment. Under vigorous wave and current action, spilt oil is found to form stable oil emulsions with mean size around 20 to 40 μ [Forrester, 1976; Berridge et al, 1968a].

The aquatic organisms are also important contributors of suspended particles. Directly, the formation of shells and skeleton of organisms, gives rise to a major part of the oceanic $CaCO_3$ precipitate [Broecker, 1976]. Diatoms are known responsible for the uptake of silicate in the marine environment [Russell-Hunter, 1970; Schink, 1967]. Indirectly, biological secretion and decomposition of organic matters can yield significant amount of suspended particles.

Man-Made Sources. Human contribution of suspended solids is related mostly to waste production. Dumping of sewage sludge and effluent from municipal or industrial wastewater treatment facilities into receiving waters has been in practice for over ten years in the United States and abroad. Between 1961 and 1972, some 389,340 m^3 of sludge mostly from the cities of Philadelphia, Pennsylvania and Camden, New Jersey were dumped some 12 miles off the mouth of the Delaware Bay [Nash, 1975]. Solid waste also finds its way to the ocean. For instance, there has been dumping of 20,000 tons and 240 tons in 1968 and 1973, respectively, of solid waste into the Pacific Ocean [Nash, 1975].

Dredging of navigatory waterways and continental drilling and construction operations are other major sources of suspended solids. Annual contribution of solids due to dredging arrives at 53 x 10^6 tons in 1968 and declines to only 37 x 10^6 tons in in 1973. Although there is a slight degree of reduction in dredging operation in most recent years, its potential of being a major contributor of suspended load still exists [Murphy and Zeigler, 1974].

The physical and Chemical Characteristics of Suspended Particles. The importance of suspended particles in influencing the distribution of chemical constituents lies in their specific physical and chemical characteristics. Proper characterization of suspended particulate is essential to insight into the mechanism of solute transport. Primarily, the mean size and size distribution, specific surface area, concentration, chemical composition, and electrokinetics are pertinent to understand their physical and chemical properties. These parameters which may interact are instrumental in the analysis of the suspended particle and its role in regulating the chemical composition of natural waters.

Size and Size Distribution. Many investigators indicate that solid particles of different size tend to retain different amounts of trace elements. The concentrations of trace metals found on various sizes of solids from the effluents of wastewater treatment plants showed a clear partition effect [Chen et al., 1974; Chen and Lockwood, 1976].

Horwitz [1974] also reported that the concentration of major metals (Fe, Mn, Al, Ti, and Mg) in the marine sediments increased with increasing grain size up to around 30 μ and then decreased, while the Ca patterns are the opposite of the others.

Particles from sanitary sewage were distributed almost equally in two size groups: > 100 μ settleable and 0.2 to 100 μ suspended [Helfgott et al., 1970, Dalrymple et al., 1975]. The secondary effluent particles from wastewater treatment plants also had a bimodal distribution pattern. The smaller fraction was estimated to be about 3 to 5 μ and about 80 or 90 μ for the larger particles. The smaller particles were approximately 40 percent to 60 percent of the total population [Tchobanoglous and Eliassen, 1970].

The size distribution of suspended particles in many estuary and coastal systems is known to be associated with the sources of particles. The size of clays from the coastal plains of Georgia and South Carolina averages 0.1 to 4 μ for kaolinite, 1 to 10 μ for illite and less than 0.1 μ for montmorillonite [Price and Calvert, 1975].

Horwitz [1974] analyzed the coastal sediments of Iceland and found bimodal and trimodal size distribution pattern. He attributed this finding to the influence of sources and deposition mechanisms.

The size distribution of suspended oceanic biological particles also followed closely those of the sewage solid and marine sediments. A multimodal distribution was observed by Sheldon et al. [1972] for particles collected from the surface and bottom waters of the Atlantic and Pacific oceans. The size spectra varies seasonally, geographically, and vertically. Krey [1967] gave a mean equivalent spherical diameter of about 5 to 6 μ at a depth of 500 m. The recent work of Gordon [1970] showed that particles smaller than 7 μ were numerically most abundant.

Surface Area. Most of the solid-solution interactions are surface reactions and are thus affected by the specific area. Banin et al [1975] measured the specific surface area and particle size of recent lake sediments. They reported the relation between these two parameters as $S = 6/100W \Sigma (P_i/d_i)$. Where S, W, P_i and d_i are specific surface area (m^2g^{-1}), density (g, cm^{-3}), fraction and size of the ith order particles, respectively. The specific surface area of the most commonly found particles is summarized in Table 1.

Table 1. Specific Surface Area of Common Suspended Particles

Solid	Specific Surface Area m^2g^{-1}
Quartz	5-10
Al_2O_3	100-200
Fe_2O_3	16-35
Calcium Carbonate	10-20
Smectite	750-800
Kaolinite	15-25
Polygorskite	450-550
Sediments (Lake Kinneret)	182±60
Organic Matter	500-750
Secondary Sludges A	450
Municipal Sludges B	75
Manganite	70

The Electrokinetic Characteristics. Near-neutral pH values, clays, most insoluble oxides, organic pollutants, bacteria, algae, and hydrocarbons are characterized by a negative surface potential as indicated by a negative electrophoretic mobility (or zeta potential). The surface charge may originate from dissociation of chemical groups, from adsorption of ionic species; in certain cases, surface imperfection and lattice substitution can also cause the charge [Stumm and Morgan, 1971].

The pH-dependent of surface charge observed in many solids is due to the fact that H^+ ion or its complexes are potential determining species. From Figure 1, it is seen that at pH values of zero electrophoretic mobility, the shearing plane across from the surface of the colloidal particle has zero net charge. This point is defined as pH_{zpc}, the point of zero charge. Table 2 gives the values of pH_{zpc} for typical suspended particles found in the aquatic system.

Faust and Manger [1964] reported the electrophoretic mobility of wastewater solids ranging from -0.55 to -3.75 (μ/sec)/(v/cm) This is in agreement with the data of Tchobanoglous and Eliassen [1970] who found -1.5 (μ/sec)/(v/cm) or (-20 mv) for secondary effluent solids. The average zeta potential of raw water particles averaged -18 mv and ranged from -15.5 to -22.6 mv [Baker et al., 1965].

When the ionic strength of a system increases, the electrophoretic mobility decreases, accordingly. Neihof and Loeb [1972] investigated the surface charge of particulate matter from seawater and reported that the electrophoretic mobility values for bentonite, calcium carbonate, and polysaccharide are -1.11, -0.50, and -0.15 (μ/sec)/(v/cm) respectively. The electrophoretic mobility declines linearly as square root of salinity as predicted.

The Chemical Composition of Suspended Particulates. Suspended particulates generally consist of inorganic and organic fractions. Depending upon the source, origin and transport mechanisms, the chemical composition varies. The inorganic portion consists of mostly quartz, feldspar, mica, clay minerals, and amorphous oxides [Neisheisel, 1973].

In the oceanic environment, the chemical composition of suspended particulates is largely determined by their origins. Detrital materials, which are derived mainly from continental erosion, generally contain quartz, orthoclase, plagioclase, kaolinite, illite, montmorillonite, and chlorite. Authigenic minerals which are formed by spontaneous crystallization are inorganic by nature. Calcite and opal are typical marine precipitates produced by floating planktons [Broecker, 1976].

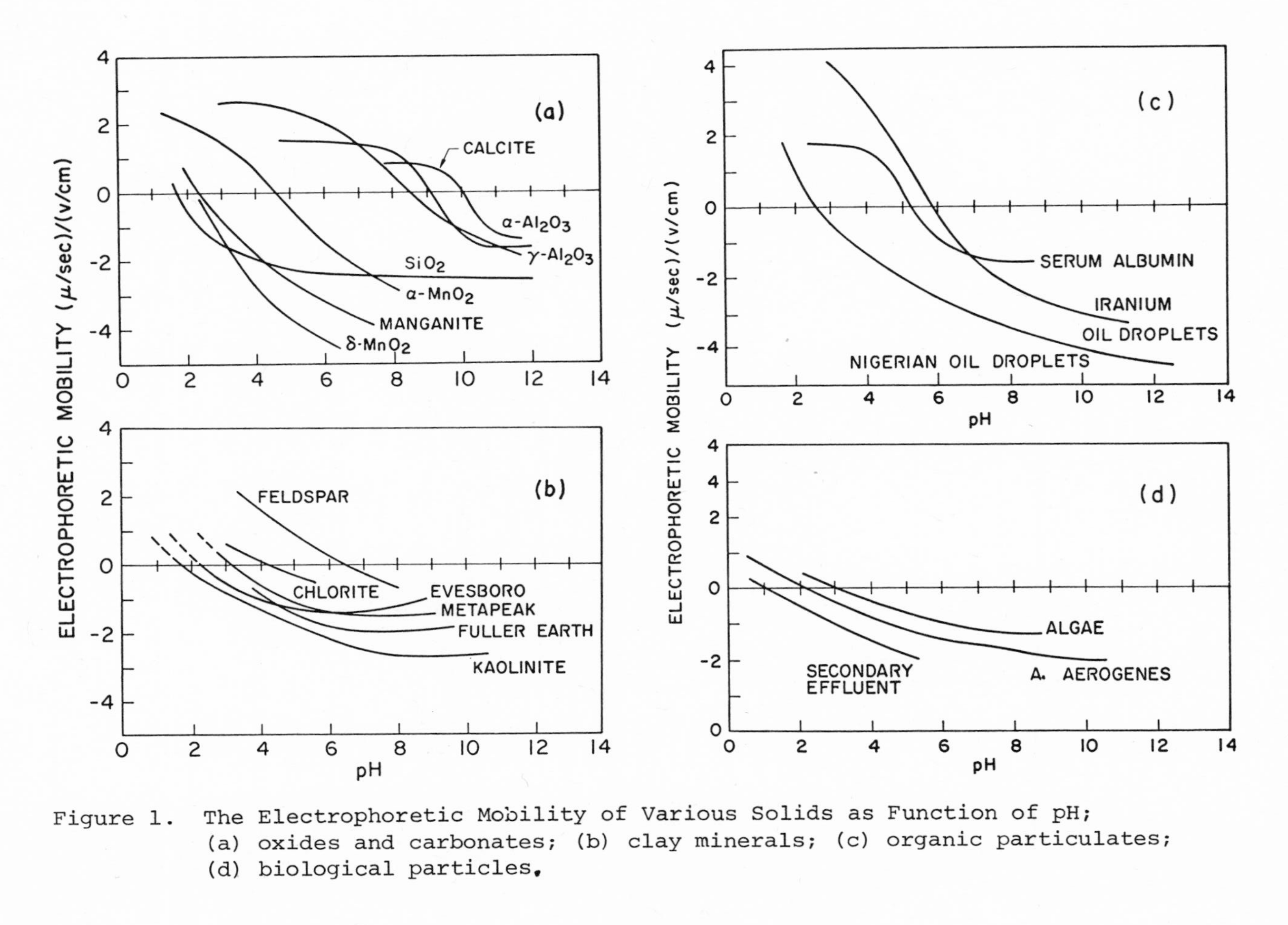

Figure 1. The Electrophoretic Mobility of Various Solids as Function of pH; (a) oxides and carbonates; (b) clay minerals; (c) organic particulates; (d) biological particles.

Table 2. The Zero Point of Charge of Commonly Found Suspended Particles

Solids	pH_{zpc}	Solids	pH_{zpc}
γ-$A\ell(OH)_3$	9.1	SiO_2	1-2
α-$A\ell OOH$	7.6	$CaCO_3$	9-10
α-$A\ell O_3$	8.4	Kaolinite	2-5
γ-$A\ell OOH$	8.4	Chloride	4.0
γ-$A\ell_2O_3$	8.5	Feldspar	6.6
β-MnO_2	7.2	Metapeak	3.2
γ-MnO_2	5.5	Evesboro	2.2
α-MnO_2	4.6	Tabaco Mosaic Virus	3.6
δ-MnO_2	1.5	A. Aerogens	2.0
		Algae	3.0

In many cases, the mineral composition of the suspended particles resemble that of the bottom sediment [Boothe and Knauer, 1972]. The mineral compositions of surface suspended solids range from a low magnesium calcitearagonite suite of Florida to montmorillonite-kaolinite combination from Alabama to Texas coastal regions [Manheim and Hathaway, 1972]. Similar composition is found for the recent marine sediments.

The concentration of suspended solids in aqueous systems varies. Table 3 summarizes the concentration of suspended particulates typical to natural waters and wastewater effluents as well. A continuing decrease in suspended solid concentration from wastewater effluents to the ocean water is clearly seen from Table 3.

Table 3. Typical Concentration of Suspended Solid Found in the Aquatic Environment

Location	mg/ℓ Concentration
Anclote Estuary	22.1–47.2
Delaware Estuary	20–405
Delaware River	65
North Atlantic Ocean	0.05–1.0
South Pacific Ocean	0.02–0.50
Atlantic-Pacific-Indian Ocean	
organic carbon (surface)	0.002–0.015
plankton (surface)	0.004–0.008
Arctic Ocean	
(organic matter)	0.009–0.055
Combined Sewer Overflow	
Total Suspended	290
Volatile Suspended	156
Settleable	151
Sanitary Sewage	
Settleable	148
Total Suspended	249
Volatile Suspended	172

ADSORPTION AT SOLID-SOLUTION INTERFACES

Removal of Metal Ions from Dilute Aqueous Solutions

Much has been reported on the removal of metal ions from dilute aqueous solution by various solids, e.g. hydrous oxides, clays, sediments, and organic detritals.

All available information indicate that pH, the chemical composition, and crystal structure of the solid and the concentration of adsorbate govern significantly the adsorption of metal ions at the solid-solution interface.

The effect of pH on the removal of metal ions, e.g. Fe(III), Cr(III), Co(II), Hg(II), Mg(II), Ca(II), Cu(II), Zn(II), Cd(II), pb(II), by SiO_2 is shown in Figure 2a and 2b [James and Healy, 1972; Huang et al., 1977]. This is attributed to the amphoteric nature and the hydrolysis behavior of metal ions. Adsorption increases abruptly at certain pH value. As pH increases, adsorption increases significantly; then chemical precipitation begins to play its removal mechanism dominantly.

Negative adsorption was observed when both solid and the absorbate are positively charged. Figures 2b and 2c show that Cd(II) and Pb(II) may be negatively adsorbed at certain pH values [Huang et al., 1977]. One can also note that in most aquatic systems with pH values ranging from 6.5 to 8.5, only Co(II) and Zn(II) are significantly removed by SiO_2. Cd(II), on the contrary, is negatively adsorbed. Ca^{+2} and Mg^{+2} ions are only slightly removed. Hg(II) is mostly removed by precipitation if no other adsorbents are present (Figures 2a and 2d).

Adsorption of trace metals on $A\ell_2O_3$ is rarely reported. Figure 2c gives adsorption of four heavy metals by γ-$A\ell_2O_3$ [Huang et al., 1977]. In the natural water environment γ-$A\ell_2O_3$ is mostly responsible for the removal of Cd(II). The absorption characteristics of metal ions onto goethite is demonstrated in Figure 2d. It shows that Zn(II), Co(II), and Cd(II) are mostly removable by goethite in the natural waters [Forbes and Quirk, 1974].

The concentration of adsorbate can also affect adsorption reaction. Loganathan [1971] reported that the percentage of Co(II) removal increases with decreasing absorbate concentration. This is due to the availability of surface sites. However, the absolute mass of metal removed does increase with increasing adsorbate concentration.

The crystal structure of absorbent determines the field strength, influences water structure, and thereby affects the adsorption reaction. The magnitude of metal ion adsorption is at least 10 to 100 times greater for δ-MnO_2 than manganite [Loganathan, 1971; Murray et al., 1969]. Stumm et al. [1970] suggested that oxides such as β-MnO_2, which have strong field strength in aqueous solution, facilitate its metal association capacity. This is in good agreement with the model proposed by James and Healy [1972]. Solids with strong field strength exhibit great free energy of solvation and free energy of chemical adsorption, and thereby enhance the extent of metal adsorption.

Soils and clay minerals are products of meteorological and biological actions which bring solid dynamics and heterogeneous nature. Hence, many chemical components are found to react with soils and their related minerals, e.g. clays. Lindsay and Norrall [1969] reported an ion-exchange process for metal-soil reaction. Two moles of protons are released from soil surface per one mole of metal ion adsorbed. Huang et al. [1977] found that the adsorption of heavy metals onto soil surface is also a pH-dependent reaction. They demonstrated that in natural environment, metal ion adsorption by soils is selective. For example, only Zn(II) and Cd(II) are removable by the two Delawarean soils (i.e. metapeak and ersboro) studied.

The metal-holding capacity of soils has been attributed to the presence of oxides, clay minerals or organics. This is evident by the correlation between size, clay content, oxide content and organic content of soils with metal adsorption capacity [Foster and Hunt, 1975]. Fordham [1969] investigated the iron (IV) adsorption onto kaolinite and concluded that adsorption is pH-dependent. The absorbed iron is chemisorption by nature and located at the edge sites on the clays. At higher pH values and prior to the precipitation of iron hydroxide, polymerization of adsorbed iron is taking place on the clay surface.

A parallel study done by Carrol [1958] revealed that iron adsorption onto clay surfaces depends on the oxide content of clay. She elaborated that iron is associated with clays in several ways: (1) as an essential constituent, (2) as a minor constituent within the crystal lattice where it is in isomorphous substitution, and (3) as iron oxide deposited on the surface at the mineral peleteles. She stated further that since kaolinite and halloysite do not possess sites within the lattice for iron, adsorption of iron onto these two clays can occur only through the last step.

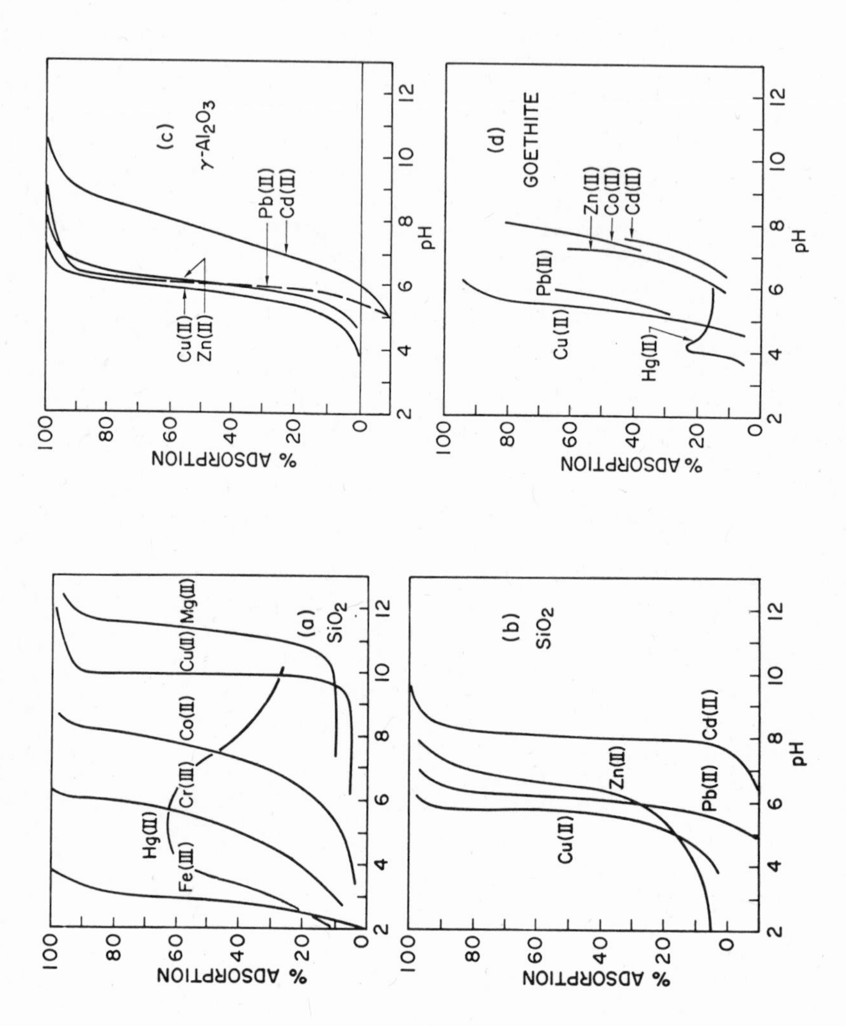

Figure 2. The Adsorption of Metal Ions by Various Solids: (a) SiO_2; (James and Healy [1972]); (b) SiO_2 (Huang et al. [1977]), (c) γ-Al_2O_3 (Huang et al. [1977]); (d) Goethite (Forbes and Quirk [1974]).

Gibbs [1973] proposed five mechanisms for metal transport in large rivers, e.g. the Amazon and Yukon Rivers. He suggested that metals are partitioned by suspended solids into five different phases: (1) in solution as free or complexed species, (2) by adsorption at solid-solution interface, (3) by incorporation into biological system, (4) by precipitation and coprecipitation as metallic coatings, and (5) by incorporation into crystalline structure. He stated that under most circumstances, solid-solution interfaces become substantial in trace metal transport in these major rivers.

In another study, Hilz et al.[1975] monitored the distribution of trace metals originated from a sewage outfall in an estuary and found that the concentration of Zn(II), Cu(II), pb(II) decrease from the point of discharge seaward. They attributed this observation to the scavanging of trace metals by adsorption, precipitation, flocculation, and biological uptake followed by deposition in the sediments.

Removal of Anions from Dilute Aqueous Solution

The removal of environmentally important anions, e.g. phosphate, silicate, arsenate, selenate, from dilute aqueous solutions is illustrated in Figure 3 [Hingston et al., 1971; Bowden et al., 1973; Hingston and Raupach, 1967; Hingston et al., 1968]. All anions are readily removed by oxides, e.g. goethite, in natural waters. In contrast to metal-oxide reaction, the adsorption density of anions increases with decreasing pH, except silicate which exhibits a maximum adsorption peak. The results indicate that the order of association increases as follows: arsenate > selenate > silicate > phosphate for goethite. Different orders are expected for different solids, since the association is determined by the properties of the solids, the solutes and their combination on the entire system [Huang, 1975a]. In typical surface waters, γ-$A\ell_2O_3$ and hematite show greater affinity toward phosphate ions than α-$A\ell_2O_3$, α-goethite. Similar results hold true for silicate adsorption Huang [1975b].

The adsorption phenomenon observed above for anions has important geological implications. Mackenzie and Garrels [1966] proposed that the small but significant x-ray amorphous alumino-silicate fraction of riverine suspended particulates may react with SiO_2, HCO_3^-, and available cations with the formation of cation alumino-silicates

$$\text{Alumino=silicate} + SiO_2(aq) + HCO_3^- + \left\{\begin{matrix} K^+ \\ \text{or} \\ Na^+ \\ \text{or} \\ Mg^{+2} \end{matrix}\right\} = \{\underset{\text{(illite)}}{K_{0.25}Al_{2.5}Si_{3.5}O_{10}(OH)_2}$$

$$\text{or } \underset{\text{(sodium montmorillonite)}}{Na_{0.33}Al_{2.33}Si_{3.67}O_{10}(OH)_2} \quad \text{or } \underset{\text{(chlorite)}}{Mg_5Al_2Si_3O_{10}(OH)_2} + CO_2 + H_2O\} +$$

$$CO_2 + H_2O$$

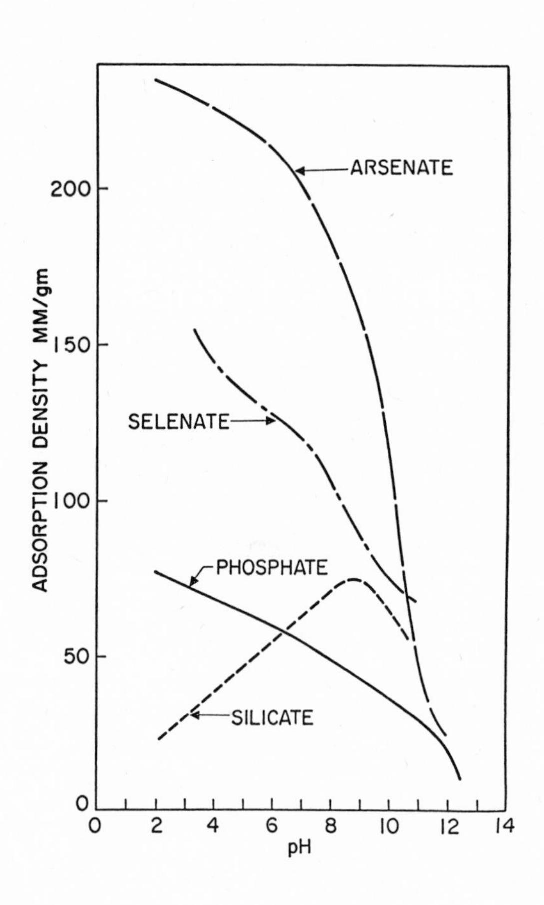

Figure 3. The Adsorption Characteristics of Anions onto Hydrous Goethite.

They further suggested that the interfacial process may serve as an important regulatory mechanism for marine silicate concentration and probably primary productivity. A similar process with describes the adsorption of silicate by suspended riverine solids was proposed by Bien et al. [1958].

Much has been reported on the uptake of phosphate anions by alumina, kaolinites, and montmorillonite [Chen et al., 1973] and by precipitated ferric and aluminum hydroxide [Carritt and Goodjal, 1954; Hsu, 1964]. Bortleson and Lee [1974] reported that the phosphate concentration in six Wisconsin lake sediments is closely correlated to iron and maganese. They further related this phosphate adsorptive and retentive capacity of sediments to the iron content. Shukla et al. [1971] found that the adsorption capacity for phosphate of natural lake sediments was greatly reduced by removing iron and aluminum hydrated oxides. Perrot et al. [1974] observed that the deposition of hydrated metal oxides onto clay surfaces tend to enhance phosphate adsorption by clay minerals. Strom [1976] reported a significant correlation between phosphate adsorbed and iron hydroxide by sediments from the Delaware Bay and the Great Marsh of Lewes.

Removal of Organics from Dilute Aqueous Solution

In natural waters, organic matters range from simple soluble molecules such as amino acids or phenols, to long-chain surfactants, to complex natural color-forming species, to macromolecules such as proteins and organic aggregates.

The removal of NTA (nitrilotriacetate) by adsorption onto calcium carbonate surfaces is an ideal example of organic acid-solid reaction. Calcite which is a well-crystalline calcium carbonate exhibits the greatest degree of adsorption density. The importance of calcium carbonate adsorption is mostly associated with marl lakes and tropical marine environment where calcium carbonate constitutes the major portion of the suspended load [Boecker, 1976].

Surface-active compounds constitute another environmentally important organic. Surfactant adsorption always increases with increasing chain length and is enhanced at low pH values, at least for the goethite which pH_{zpc} is 8.5 [Akers and Riley, 1974; Wakamatsu and Fuerstenau, 1968]. The longer the chain length, the greater the amount of surfactant adsorbed [Wayman, 1967]. The effect of clay structure on surfactant adsorption is studied by Hemwall [1963] and Hamaker et al. [1966]. Montmorillonite is more effective in removing ABS (alkylbenzene sulfonate) than kaolinite. Multivalent cation montmorillonite is superior over monovalent cation montmorillonite in surfactant adsorption.

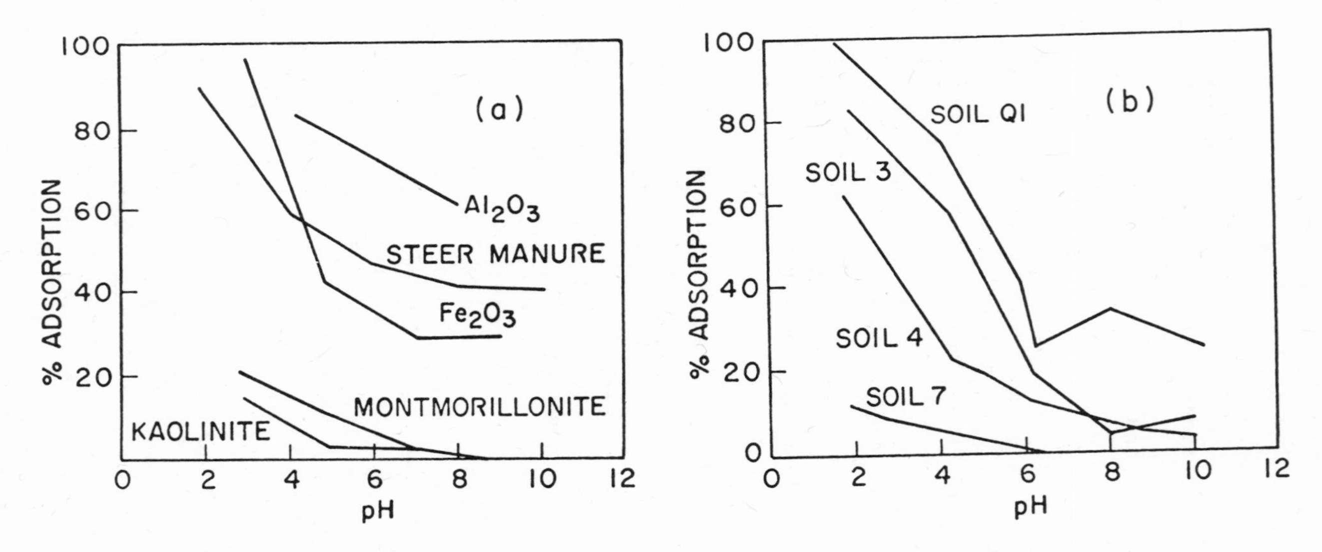

Figure 4. The Adsorption of Pesticides (4-amino-3,5,6-trichloropicolinic acid) onto Various solids: (a) oxides, clays and steer manure; (b) soils.

What remains to be considered about organic matter in the environment is pesticides. The fate of pesticides in soil-water systems was most intensively studied. Figure 4 indicates that amorphous oxides such as $A\ell_2O_3$ and Fe_2O_3 and probably organic particulates are very capable of removing pesticides in neutral pH environment. The clay minerals usually have less adsorption capacity than the above solids [Hamaker et al., 1966].

The removal efficiency of particles by soil materials varies. This is partly due to the different chemical background of the soils. Soils contain different amounts of oxides and organic matters can have different pesticide adsorption capacities. Lotse et al.[1968] studied the adsorption of lindane onto riverine sediment and found the amounts of lindane adsorbed varied from 0.7, 0.55, 0.28, 0.17 μg per mg dried sediment at an organic content of 64%, 31%, 13% and 3% respectively.

THE EFFECTS OF INTERFACIAL ADSORPTION

The Stability of Suspended Solids

Coagulation of particles is a two-step process: particle destabilization and particle transport. The destabilization step is dealing with the reduction of surface potential and elimination of energy barrier to facilitate interparticle contact and attachment. Particle transport is brought by Brownian motion (perikinetic flocculation) or velocity gradients (orthokinetic flocculation), or differential settling.

Naturally occurring colloids can be destabilized by three distinct mechanisms: (1) reduction of repulsion potential between colloids by compaction of the electrical double layer, (2) compensation of the surface charge by specific adsorption of oppositely charged species (adsorption coagulation), and (3) bridge-formation between particles by long chain polymers or polymeric species.

According to the electrostatic mechanism, the extent of coagulation is in inverse proportion to the valence of electrolyte to the sixth power (Schulze-Hardy Rule). Derjaguin-Landau and Verwey-Overbeck have independently developed a quantitative theory — the DLVO theory — to predict the effect of inert electrolyte on the stability of lyophobic colloids. The DLVO theory also predicts that the degree of double layer compression is enhanced with ionic strength and the charge of the counter ions. Values of relative stability are determined by the difference in potential of attraction and repulsion. The maximum resultant potential, Vm (erg cm^{-2}), is directly related to the easiness of aggregation.

The stability ratio, α, is defined as $\alpha = 2\kappa a \exp(Vm/KT)$; where κ, a, and KT are, respectively, the reciprocal thickness of the double layer (cm^{-1}), the radius of colloid (cm), the product of Boltzman's constant and absolute temperature which is an expression of thermal fluctuation energy level. The stability ratio will assume a zero value for extreme stable suspension and unity for complete destabilization [Hahn and Stumm, 1968]. According to the DLVO model, the stability of a suspended particle depends on particle size, surface charge, ionic strength, valence of electrolyte, and the chemical composition of the suspended particles in aqueous medium.

The reduction of surface charge, $\Delta\sigma$ ($coul/cm^{-2}$), due to specific adsorption is computed by the express: $\Delta\sigma = \Sigma F\, Z_i \Gamma_i$; where F, Z_i, and Γ_i are the faraday, charge of adsorbed species, and surface excess of the absorbate (mole cm^{-2}) respectively. The surface charge is reduced before the point of complete compensation of charge, pH_{zpc}. Charge reversal takes place after the point of zero charge has been passed. As a result of the over compensation of surface charge, restabilization of colloid occurs [Stumm and O'Melia, 1968].

The transport of colloidal materials is governed by the size of particles, the concentration of particles, the velocity gradient, and the temperature of the dispersion system. If the colloids are very large ($\simeq >1\mu$), the efficiency of particle transport is mainly a function of velocity gradient. For monodispersed suspensions, the resulting rate of particle removal corresponds to a first-order equation

$$-dN/dt = \alpha_o (4G/\Pi)\phi N$$

where α_o, G, ϕ, and N stand for collision coefficient, velocity gradient (sec^{-1}), solid concentration (v/v), and number of particles per volume, respectively. If the particles are small ($\simeq <1\mu$) and thermal motion supercedes the velocity gradient, the rate of particle transport follows a second-order reaction form.

$$-dN/dt = \alpha_p (4KT/3\mu) N^2$$

where α_p, KT, and μ represent collision coefficient, product of Boltzman constant and absolute temperature and absolute viscosity individually. Hahn and Stumm [1968; 1970] have investigated the kinetics of coagulation with prototype particles, SiO_2. Laboratory results have verified what was predicted by these equations.

The Distribution of Clay Minerals in the Estuarine Sediments

The abundance of one kind of clay over the others in the aquatic sediments can be interpreted in various ways. In estuarine sediments, the distribution of clay fractions are considered closely related to the stability of the individual clay. Based upon the colloid stability theory presented above, Edzwald et al. [1974] computed the stability of clay in the estuarines and concluded that illite is more stable than kaolinite which is more stable than montmorillonite. Hahn and Stumm [1970] also computed the stability of clay minerals and predicted that in estuarine environments, illite is coagulated after montmorillonite is in the estuary. Therefore, the least stable clays are distributed more abundantly toward the fresh water end of the estuary, while most stable clays tend to find their abundance in the salt tide zone. Regardless of the estuary system, the sedimentary illite increases seaward, while montmorillonite always decreases toward river mouths. From available data, it is clear that differential coagulation of clay minerals in an estuary contributes significantly to the distribution pattern of clays.

Distribution of Heavy Metals in the Estuarine Sediments

The importance of sediments in cycling of heavy metals in an aqueous environment has been discussed by a number of researchers [McLerran and Holmes, 1974; Rao and Rao, 1973; Khalid et al., 1975; Taylor, 1974; Chester and Stoner, 1975]. Many mechanisms are proposed: (1) selective adsorption onto oxides, clays, or aquatic bacteria followed by differential coagulation of the suspended solids and eventually arriving at the sediment; (2) chemical precipitation leading to phase transition.

Data on the absorption of heavy metal by suspended solids demonstrate the importance of oxides, specifically ferric oxides, aluminum oxides, and manganese oxides, in controlling the trace metal concentration of aquatic systems. The distribution of eight metals, e.g. Cd(II), Hg(II), pb(II), Cu(II), Zn(II), Cr(III), Ni(II), and Mg(II), in the sediments of Delaware Estuary is significantly correlated to total iron [Bopp, 1973]. The importance of ferric oxide in regulating the concentration of chemical constituents in the aquatic system goes beyond heavy metals. An excellent correlation between tied phosphorus and iron content for sediments collected from the Delaware Estuary was also observed. It is conspicuous that coagulation and deposition of ferric colloids in the upper reaches of the estuaries produce a marked decrease in the concentration of sedimentary phosphorus in a seaward direction [Strom, 1976].

SUMMARY

The absorption characteristics of soluble chemicals, e.g. trace metals, pesticides, and inorganic anions onto various solids, e.g. oxides, carbonates, clays, soils, and organic detritus exactly reflect the importance of solid-solution interface and its role in controlling the chemical composition of the natural waters.

Suspended solids are introduced into the natural bodies of water by natural forces, e.g. weathering, hydrothermal activities, biosynthesis, and volcanic emanation,or by man-made activities, e.g. sewage discharge, dredging, continental drilling, and agricultural excavation. Regardless of their sources and origins, suspended solids can be classified as inorganic oxides, carbonates, clays, and as organic detritus. The importance of suspended particles lies in their unique characters, specifically small size, large specific surface area, and charge-carrying in the aqueous systems. The concentration of suspended solids in the aquatic environment varies from a few hundredths of a mg per liter in sewage effluent, to a few tenths of a mg per liter in rivers and estuaries, to a few tenths and hundredths of a mg per liter in the oceanic environment. Major solids found in the aquatic environment are quartz, ferric oxides, aluminum oxides, clays, and organic aggregates. At the pH and the chemical conditions of most aquatic environments, most suspended solids are negatively charged.

Selective adsorption of chemical species onto suspended particles followed by differential coagulation of the suspended particulate complete the most important part of mass transport. This is partly exemplified by the trace metal distribution and the clay fractionation in the estuarine sediments.

ACKNOWLEDGEMENT

This work was supported in part by a grant, ENG75-07176, provided by the National Science Foundation.

REFERENCES

Akers, R. J. and Riley, P. W., The Adsorption of Polyoxyethylene Alkyphenol onto Calcium Carbonate from Aqueous Solution, Jour. Colloid Interface Sci., 48, 161, 1974.

Baker, R. A., Campbell, S. J., and Anspach, F. R., Electrophoretic and Taste and Odor Measurement in Plant Control, Jour. Amer. Water Works Assoc., 57, 363, 1965.

Banin, A., Gal, M., Zohar, Y., and Singer, A., The Specific Surface Area of Clays in Lake Sediments--Measurement and Analysis of Contributors in Lake Kinneret, Israel, Amer. Soc. Limno. and Oceanog., 20, 278, 1975.

Berridge, S. A., Dean, R. A., Fallows, R. G., and Fish, A., The Properties of Persistent Oils at Sea, Jour. Inst. Pet., 54, 300, 1968a.

Bien, G. S., Contors, D. E., and Thomas, W. E., The Removal of Soluble Silica from Fresh Water Entering the Sea, Geochim, et Cosmochim. Acta, 14, 35, 1958.

Boothe, P. N. and Knauer, G. A., The Possible Importance of Fecal Material in the Biological Amplification of Trace and Heavy Metals, Amer. Soc. Limno. and Oceanog., 17(2), 270, 1972.

Bopp, F., III, The Baseline Concentration of Trace Metal in Delaware Estuary, M. S. Thesis, University of Delaware, Newark, Delaware, 1973.

Bortleson, G. C. and Lee, G. F., Phosphorus, Iron, and Manganese Distribution in Sediment Cores of Six Wisconsin Lakes, Amer. Soc. Limno. and Oceanog., 19, 794, 1974.

Bowden, J. N., Bolland, M. D. A., Posner, A. M., and Quirk, J. P., Generalized Model for Anion and Cation Adsorption at Oxide Surfaces, Nature, 245, 81, 1973.

Broecker, W. S., Chemical Oceanography, Harcourt Brace Jovanovich, Inc., 1976.

Carritt, D. E. and Goodjal, S., Sorption Reactions and Some Ecological Implications, Deep Sea Res., 1, 224, 1954.

Carroll, D., Role of Clay Minerals in the Transportation of Iron, Geochim. et Cosmochim. Acta, 14, 1, 1958.

Chen, K. Y., Young, C. S., Jan, T. K., and Rohatgi, N., Suspended and Dissolved Trace Metals in Wastewater Effluents, Jour. Water Poll. Control Fed., 6, 2663, 1974.

Chen, K. Y. and Lockwood, R. A., Evaluation Strategies of Metal Pollution in Oceans, Jour. Environ. Div., Proceedings of Amer. Soc. Civil Eng., 102(EE2), 347, 1976.

Chen, Y. S., Butler, J. N., and Stumm, W., Kinetics Study of Phosphate Reaction with Aluminum Oxide and Kaolinite, Environ. Sci. and Technol., 4(4), 327, 1973.

Chester, R. and Stoner, J. H., Trace Elements in Total Particulate Material from Surface Sea Water, Nature, 255, 50, 1975.

Dalrymple, R. J., Hodd, S. L., and Morin, D. C., Physical and Settling Characteristics of Particulates in Storm and Sanitary Wastewaters, EPA-670/2-75-011, 1975.

Edzwald, J. K., Upchurch, J. B., and O'Melia, C. R., Coagulation in Estuaries, Environ. Sci. and Technol., 8, 58, 1974.
Faust, S. D. and Magner, M. C., Electromobility Values of Particulate Matter in Domestic Wastewater, Water and Sewage Works, 111, 1964.
Forbes, E. A. and Quirk, J. P., The Specific Adsorption of Inorganic Hg(II) Species and Co(III) Complex Ions on Goethite, Jour. Colloid Interface Sci., 49(3), 403, 1974.
Fordham, A. N., Sorption and Precipitation of Iron on Kaolinite, II. Sorption Isotherms and the Interpretation in Terms of Iron (III) Ionic Equilibria, Australian Jour. Soil Res., 7, 199, 1969.
Forrester, W. D., Distribution of Suspended Soil Particles Following the Grounding of the Tanker Arrow, Jour. Marine Res., 29, 151, 1976.
Foster, P. and Hunt, D. T. E., Geochemistry of Surface Sediments in an Acid Stream Estuary, Marine Geol., 18, M13-21, 1975.
Gibbs, R. J., Mechanisms of Trace Metal Transport in Rivers, Science, 180(4081), 71, 1973.
Gordon, D. C., A Microscopic Study of Organic Particles in the North Atlantic Ocean, Deep Sea Res., 17, 175, 1970.
Graham, W. S., The Adsorption Characteristics of Free Amino Acids and Protein at the Solid-Solution Interface, M. S. Thesis, University of Delaware, Newark, Delaware, 1976.
Hahn, H. H. and Stumm, W., The Role of Coagulation in Natural Waters, Amer. Jour. Sci., 268, 354, 1970.
Hamaker, J. W., Goring, C. A. I., and Youngson, C. R., Sorption and Leaching of 4-Amino-3,5.6-Trichloropicolinic Acid in Soils, in Organic Pesticides in the Environment, Adv. Chem. Ser. 60, A.C.S., 280, 1966.
Helfgott, T., Hunter, J. V., and Rickert, D., Analytical and Process Classification of Effluents, Jour. Sanitary Eng. Div., Proceedings of Amer. Soc. Civil Eng., SA3, 779, 1970.
Helz, G. R., Huggett, R. J., and Hill, J. M., Behavior of Mn, Fe, Cu, Zn, Cd, and pb Discharged from a Wastewater Treatment Plant into an Estuarine Environment, Water Res., 9, 631, 1975.
Hemwall, J. R., The Removal of 4-tert-butylpyrocatechol (TBC) by Clay Minerals, Proceedings Intern. Clay Mineral Conf., Stockholm, 1, 319, 1963.
Hingston, F. J. and Raupach, M., The Reaction Between Monosilic Acid and Aluminum Hydroxide, I. Kinetics of Adsorption of Silic Acid by Aluminum Hydroxide, Australian Jour. Sci., 51, 295, 1967.
Hingston, F. J., Posner, A. M., and Quirk, J. P., Competitive Adsorption of Negatively Charged Ligands on Oxide Surfaces, Discussion of the Faraday Society, No. 52, 334, 1971.
Hingston, F. J., Atkinson, R. J., Posner, A. M., and Quirk, J. P., Specific Adsorption of Anions on Goethite, Ninth International Conference, Soil Sci. Adelaide, 1, 669, 1968.
Horowitz, A., The Geochemistry of Sediments from the North Reycjanes Ridge and the Iceland-Faroes Ridge, Marine Geol., 17, 103, 1974.
Hsu, P. H., Adsorption of Phosphate by Aluminum and Iron in Soils, Soil Sci. Soc. Amer. Proc., 28, 174, 1964.

Huang, C. P., The Adsorption of Phosphate at the Hydrous γ-$A\ell_2O_3$ Electrolyte Interface, Jour. Colloid Interface Sci., 53, 178, 1975a.

Huang, C. P., The Removal of Aqueous Silica from Dilute Aqueous Solution, Earth and Planetary Sci. Letters, 27, 205, 1975b.

Huang, C. P., Elliott, H. A., and Ashmead, R. A., Interfacial Reactions and the Fate of Heavy Metals in Soil-Water Systems, J. Water Pollution Control Federation, 49, 745, 1977.

James, R. O. and Healy, T. W., Adsorption of Hydrolyzable Metal Ions at Oxide-Water Interface, Jour. Colloid Interface Sci., 40(1), 65, 1972.

Khalid, R. A., Gambrell, R. P., and Patrick, W. H., Jr., Sorption and Release of Mercury by Mississippi River Sediment as Affected by pH and Redox Potential, Biological Implications of Metals in the Environment, 15th Ann. Life Sci. Symp., Richland, Washington, 1, 1975.

Krey, J., Detritus in the Ocean and Adjacent Sea, in Estuarines (ed., G. H. Lauff), Amer. Assoc. Advan. Sci. Pub., 83, 1967.

Krumbein, W. C. and Garrels, R. M., Origin and Classification of Chemical Sediments in Terms of pH and Oxidation-Reduction Potential, Jour. Geol., 60, 1, 1952.

Leckie, J. O. and James, R. O., Control Mechanisms for Trace Metals in Natural Waters, in Aqueous-Environmental Chemistry of Metals (ed., A. J. Rubin), 1, 1974.

Lindsay, W. L. and Norvall, W. A., Equilibrium Relationships of Zn^{+2}, Fe^{+2}, Ca^{+2}, and H^+ with EPTA and DTPA in Soils, Soil Sci. Soc. Amer. Proc., 33, 62, 1969.

Loganathan, P., Sorption of Heavy Metals in a Hydrous Manganese Oxide, Ph.D. Thesis, University of California at Davis, 1971.

Lotse, E. G. Graetz, D. A., Chesters, G., Lee, G. B. and Newland, L. W., Lindane Adsorption by Lake Sediments, Jour. Environ. Sci. and Technol., 2, 353, 1968.

Lush, D. L. and Hynes, H. B. N., The Formation of Particles in Freshwater Leachates of Dead Leaves, Amer. Soc. Limno. and Oceanog., 18, 968, 1973.

Mackenzie, F. T. and Garrels, R., Silicate-Bicarbonate Balance in the Ocean and Early Diagnosis, Jour. Sedimentary Petrology, 36(4), 1075, 1966.

Manheim, F. T. and Hathaway, J. C., Suspended Matter in Surface Waters of the Northern Gulf of Mexico, Amer. Soc. Limno. and Oceanog., 17(1), 17, 1972.

Mason, B., Principles of Geochemistry, John Wiley and Sons, New York, Third Edition, 1966.

McLerran, C. J. and Holmes, C. W. Deposition of Zinc and Cadmium by Marine Bacteria in Estuarine Sediment, Amer. Soc. Limno. and Oceanog., 19, 998, 1974.

Morgan, J. J. and Stumm, W., The Role of Multivalent Metal Oxides in Limnological Transformations, as Exemplified by Iron and Manganese, Proceedings of Second International Water Poll. Res. Conf., 103, 1964.

Murray, D. J., Healy, T. W., and Fuerstenau, D. W., The Adsorption of Aqueous Metal on Colloidal Hydrous Manganese Oxide in Adsorption from Aqueous Solution, Adv. Chem. Ser., 79, A.C.S., 1969.

Murphy, W. L. and Zeigler, T. W., Practices and Problems in the Confinement of Dredged Material in Corps of Engineers Project, U. S. Army Engineers Waterways Experiment Station, Vicksburg, Mississippi, Tech. Rept. D-74-2, 1974.

Nash, N., Sludge Disposal and the Coastal Metropolis (ed., T. Church), Marine Chemistry in the Coastal Environment, A.C.S. Symposium Ser. 18, 1975.

Neihof, R. A. and Loeb, G. I., The Surface Charge of Particulate Matter in Seawater, Amer. Soc. Limno. and Oceanog., 17, 7, 1972.

Neisheisel, J., Long Range Spoil Disposal Study, Part III. Sub-study of Nature, Source, and Cause of the Shoal, Appendix A, U. S. Army Engineering District, Corps of Engineers, 1, 1973.

Perrot, K. W., Langdon, A. G., and Wilson, A. T., Sorption of Anions by the Cation Exchange Surface of Muscovite, Jour. Colloid Interface Sci., 48, 10, 1974.

Price, N. B. and Calvert, S. E., A Study of the Geochemistry of Suspended Particulate Matter in Coastal Waters, Marine Chem., 1, 169, 1975.

Rao, N. V. N. D. and Rao, M. P., Trace Element Distribution in the Continental Shelf Sediments Off the East Coast of India, Marine Geol., M43, 1973.

Riley, G. A., Organic Aggregates in Seawater and the Dynamics of Their Formation and Utilization, Limno. Oceanog., 8, 372, 1963.

Riley, G. A., Particulate Organic Matter in Seawater, Adv. Mar. Biol., 8, 1, 1970.

Russell-Hunter, W. D., Aquatic Productivity, Macmillan Co., Collier-Macmillan, Ltd., London, 30, 1970.

Schink, D. R., Budget for Dissolved Silica in the Mediterranean Sea, Geochim. et Cosmochim. Acta, 31, 987, 1967.

Sheldon, R. W., Prakash, A., and Sutcliffe, W. H., Jr., The Size Distribution of Particles in the Ocean, Amer. Soc. Limno. and Oceanog., 17, 327, 1972.

Sheldon, R. W., Sutcliffe, Jr., W. H., and Prakash, A., The Production of Particles in the Surface Waters of the Ocean with Particular Reference to the Sargasso Sea, Amer. Soc. Limno. and Oceanog., 18(5), 719, 1973.

Shukla, S. S., Syers, J. K., Williams, J. D. H., Armstrong, D. E., and Harris, R. I., Sorption of Inorganic Phosphate by Lake Sediments, Proc. Soil Sci. Soc. Am., 35, 244, 1971.

Strom, R. N., Phosphorus Fractionation in Estuary and Marsh Sediments, Ph.D. Thesis, University of Delaware, Newark, Delaware, 1976.

Stumm, W., Huang, C. P., and Jenkins, S. R., Specific Chemical Interaction Affecting the Stability of Dispersed Systems, Croatica Chemica Acta, 42, 223, 1970.

Stumm, W., and O'Melia, C. R., Stoichiometry of Coagulation, Jour. Amer. Water Works Assoc., 60(5), 514, 1968.

Stumm, W., and Leckie, J. O., Phosphate Exchange with Sediments; Its Role in the Productivity of Surface Waters, Fifth International Water Poll. Res. Conf., Pergammon Press, Ltd., 1971.

Stumm, W. and Morgan, J. J., Aquatic Chemistry, Chapter 9, John Wiley & Sons, New York, 1970.

Taylor, D., National Distribution of Trace Metals in Sediments from a Coastal Environment, Tor Bay, England, Estuarine and Coastal Marine Sci., 2, 417, 1974.

Tchobanoglous, G. and Eliassen, R., Filtration of Treated Sewage Effluent, Jour. Sanitary Eng. Div., Proceedings of Amer. Soc. Civil Eng., SA2, 243, 1970.

Turekian, K. K., Oceans, Prentice-Hall, New York, 18, 1968.

Wakamatsu, T. and Fuerstenau, D. W., The Effect of Hydrocarbon Chain Length of the Adsorption of Sulfonates at the Solid/Water Interface in Adsorption from Aqueous Solution, Adv. Chem. Ser. No. 79, A.C.S., 1968.

Wayman, C. H., Adsorption on Clay Mineral Surface, in Principles and Applications of Water Chemistry (ed., Faust, S. D. and Hunter, J. V.), 127, 1967.

TRANSVERSE CURRENTS IN THE ST. LAWRENCE ESTUARY: A THEORETICAL TREATMENT

T.S. Murty and M.I. El-Sabh

Marine Environmental Data Service, Ocean and Aquatic Sciences; Section d'Océanographie, Université du Québec à Rimouski

ABSTRACT

Frequent transverse currents have been observed in the middle of the lower St. Lawrence estuary, which are directed from the south towards the north shore with a mean residual velocity of 0.6 knots in the upper layer. The explanation that is invoked for the existence of these currents follows along the lines of the classical Rossby Adjustment Problem. The mutual adjustment of pressure and velocity fields in a fluid to which momentum has been imparted (for example by winds) takes place in a few hours in the homogeneous case but occupies several days (7 to 10) in the stratified case. Since the weather systems usually travel over the St. Lawrence estuary with roughly the same frequency (as the duration of the adjustment in the baroclinic situation) and impart momentum, continuity and frequent occurrence are achieved by the transverse currents.

INTRODUCTION

The St. Lawrence estuary is a water body in eastern Canada (Figure 1) that joins the St. Lawrence River with the Gulf of St. Lawrence. Historically, the estuary formed the first section of the navigation routes which led to the settlement and development of much of Canada and the central USA. To-day, the estuary is still a vital artery to these areas forming a navigable waterway from the Atlantic Ocean to the international Great Lakes. In addition, many industries are located along its southern and northern shores. For understanding the spread and dispersion of pol-

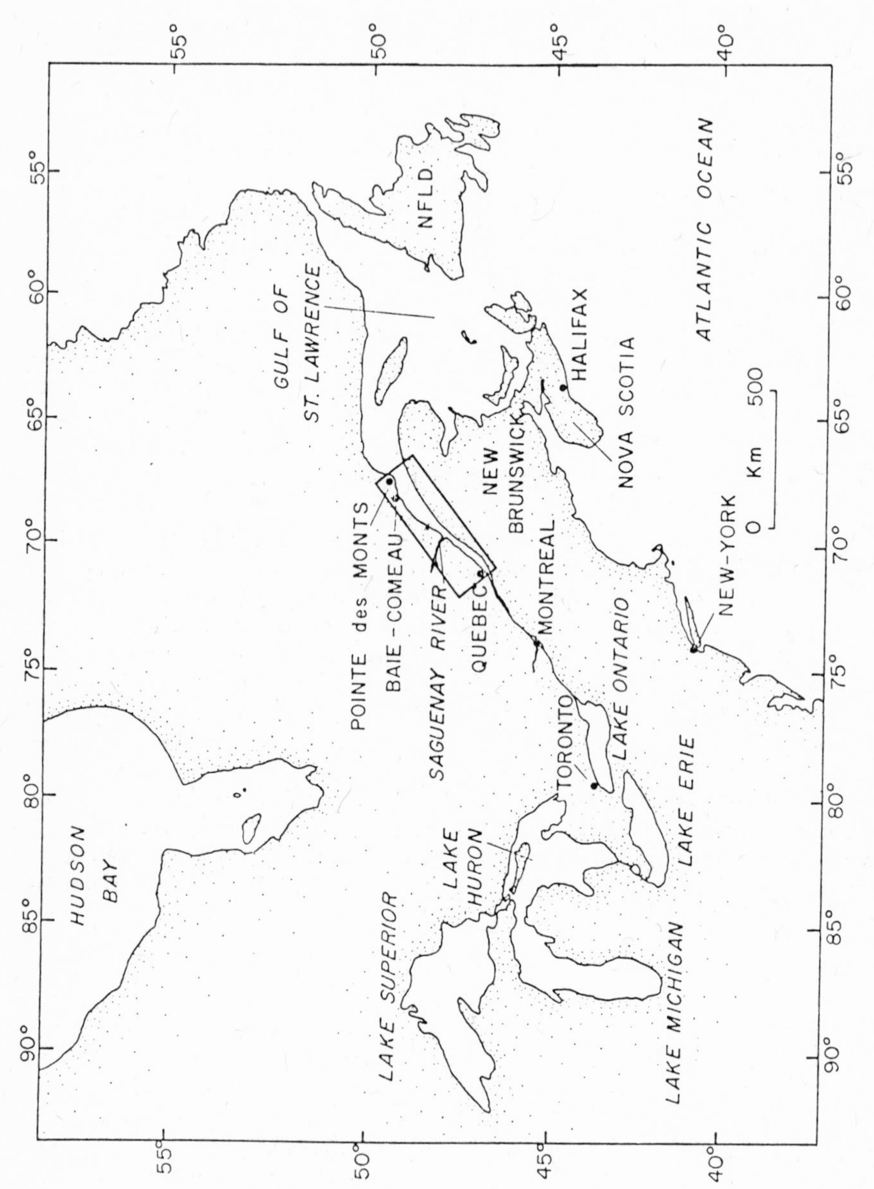

Fig. 1. General map showing the International Great Lakes, St. Lawrence River, St. Lawrence estuary and Gulf, and eastern Canada.

lutants in this estuary, a knowledge of the circulation pattern and currents is essential.

The estuary is funnel shaped with an area of 11000 km^2 and begins at the upper limit of the salt water intrusion near Quebec City, and may be extended 400 km downstream to Pointe-des-Monts, where its channel suddenly opens up into the Gulf (Figure 2). Near Pointe-des-Monts the estuary width is approximately 100 km whereas upstream of Quebec City, it narrows to about a mile (1.6 km) width. A natural subdivision into an upper and lower estuary is suggested by the bottom physiography. A major break-in-slope and change in regional depths along the axis of the channel occurs near the confluence of the Saguenay River and the upper estuary, where the bottom rises drastically from a depth of approximately 350 m to 25 m over a 16 km distance.

With a mean discharge of 10000 m^3/s at Quebec City, the St. Lawrence River constitutes the largest single source of fresh-water input. All other major fresh-water inflows, the Saguenay, Bersimis, Outardes and Manicouagan Rivers, enter the system from the north shore, the inflow from the south shore being insignificant. The currents in the estuary exhibit several time and space scales and the observed circulation pattern is a result of the combined action of several factors such as winds, atmospheric pressure gradients, fresh-water discharge, stratification, tides, earth's rotation, friction, internal waves and topography.

Although the currents in the St. Lawrence estuary have been investigated as early as in 1892 (Farquharson, 1966), much still remains unknown about the mechanisms involved in the circulation pattern in the estuary. Here we will specifically concern ourselves with the transverse currents observed in the lower estuary, near Rimouski, which are directed from the south to the north shore. Although, they may vary in intensity these are of frequent occurrence. The explanation invoked here is that wind stress interacting with the topography combined with the influence of stratification and earth's rotation will account for these transverse currents. This problem will be qualitatively examined here following the concepts of the classical "Rossby Adjustment Problem" which deals with the mutual adjustment of pressure and velocity fields in a fluid to which momentum has been imparted.

In the homogeneous case the adjustment takes place within a few hours and hence the duration of the existence of the transverse currents has to be brief. However, in the stratified case, the adjustment takes place over several days (say 7 to 10 days). It so happens that this is also roughly the frequency with which the weather systems travel at these latitudes and thus they impart momentum to the water body at periodic intervals. Thus, although the transverse currents arising out of the adjustment process

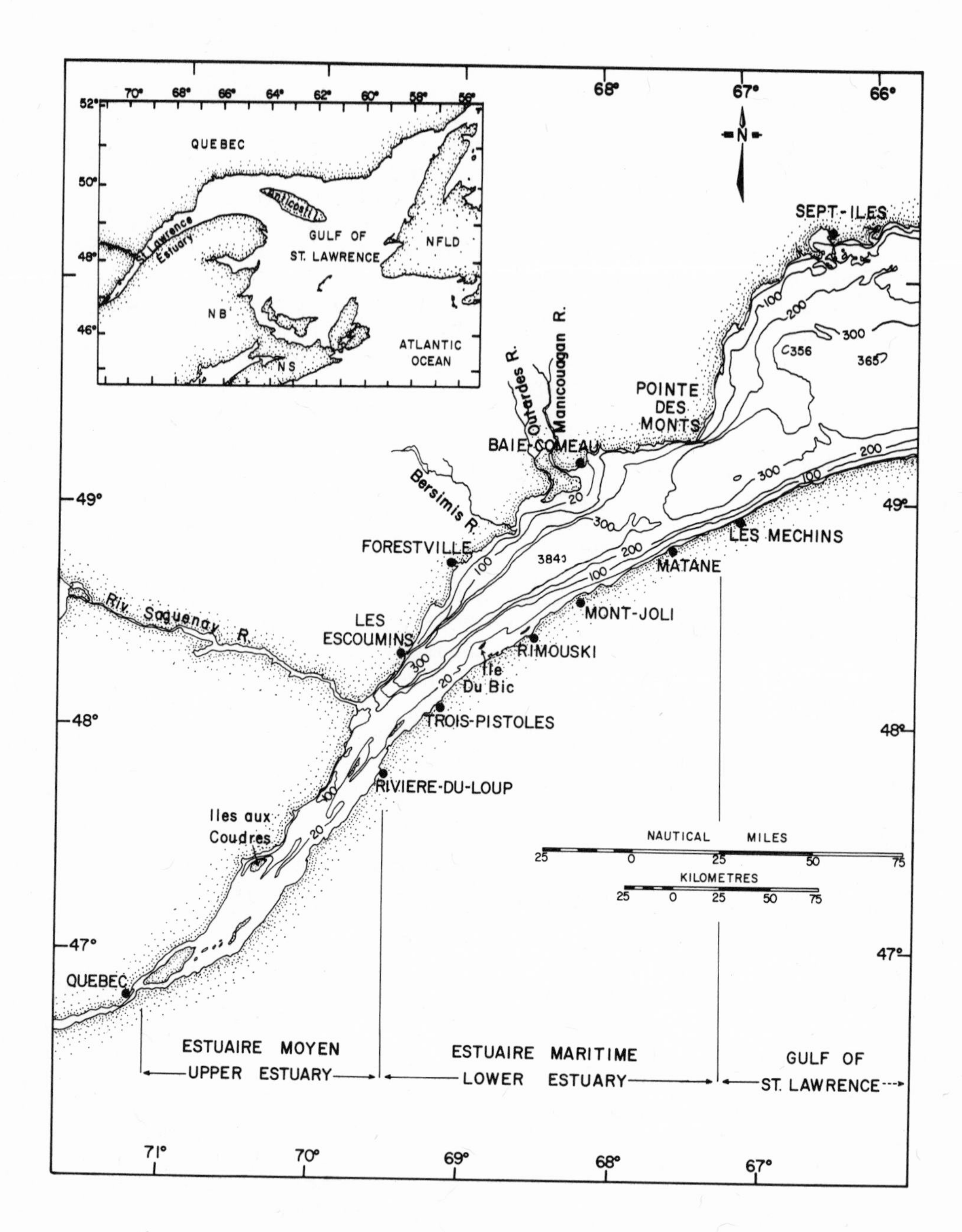

Fig. 2. Bathymetric chart for the St. Lawrence estuary. Figures indicate the depth in meters.

after momentum has been imparted from one single weather system is a transient phenomenon and may last a maximum of about 10 days, it is the passage of a succession of weather systems that give rise to the continuity and frequent occurrence of the transverse currents.

METEOROLOGICAL ASPECTS

It is of interest to briefly look at the nature of the weather systems that affect North America in general and the region of the St. Lawrence estuary in particular. Most of the Pacific cyclones cannot cross the Rocky Mountains; however, some of them redevelop on the eastern side of the Rockies. There are three areas where such a redevelopment occurs frequently (a) the region east of Sierra-Nevada; the cyclones generated here are usually weak, (b) east of the Rockies in Colorado; many of these cyclones originating here achieve great intensities and travel to the central and eastern parts of North America. They usually travel northeastward toward the Great Lakes, (c) east of the Canadian Rockies in Alberta and these storms are also intense.

During winter, the Great Lakes region is also an area of high cyclone frequency. There are mainly three tracks of depressions (Figure 3a): (a) depressions from the west move eastward between 45° and 50° N, (b) some depressions first travel southeastward till central part of USA and then travel northeastward towards New England and the Gulf of St. Lawrence and (c) depressions form on polar front off the east coast of USA and move northeastwards towards Newfoundland.

In the summer period, the frequency of east coast-originated depressions is less and the tracks of depressions from the west are somewhat northward as compared to their winter positions (Figure 3b). The tracks pass over Hudson Bay, Labrador or the Gulf of St. Lawrence.

TRANSVERSE CURRENTS IN THE LOWER ST. LAWRENCE ESTUARY: OBSERVATIONS

Current observations in the St. Lawrence estuary are scarce with poor spatial coverage. On occasion, direct current measurements were carried out at Pointe-des-Monts cross-section (Farquharson, 1966), at Rimouski cross-section (Forrester, 1967, 1970), at locations along the axis of the estuary between Trois-Pistoles and Baie-Comeau (Forrester, 1974) and near Les Escoumins (Ingram, 1975) for periods varying between two weeks and two months.

Farquharson (1966) suggests that there is an anti-cyclonic eddy in the lower St. Lawrence estuary, with its centre somewhere

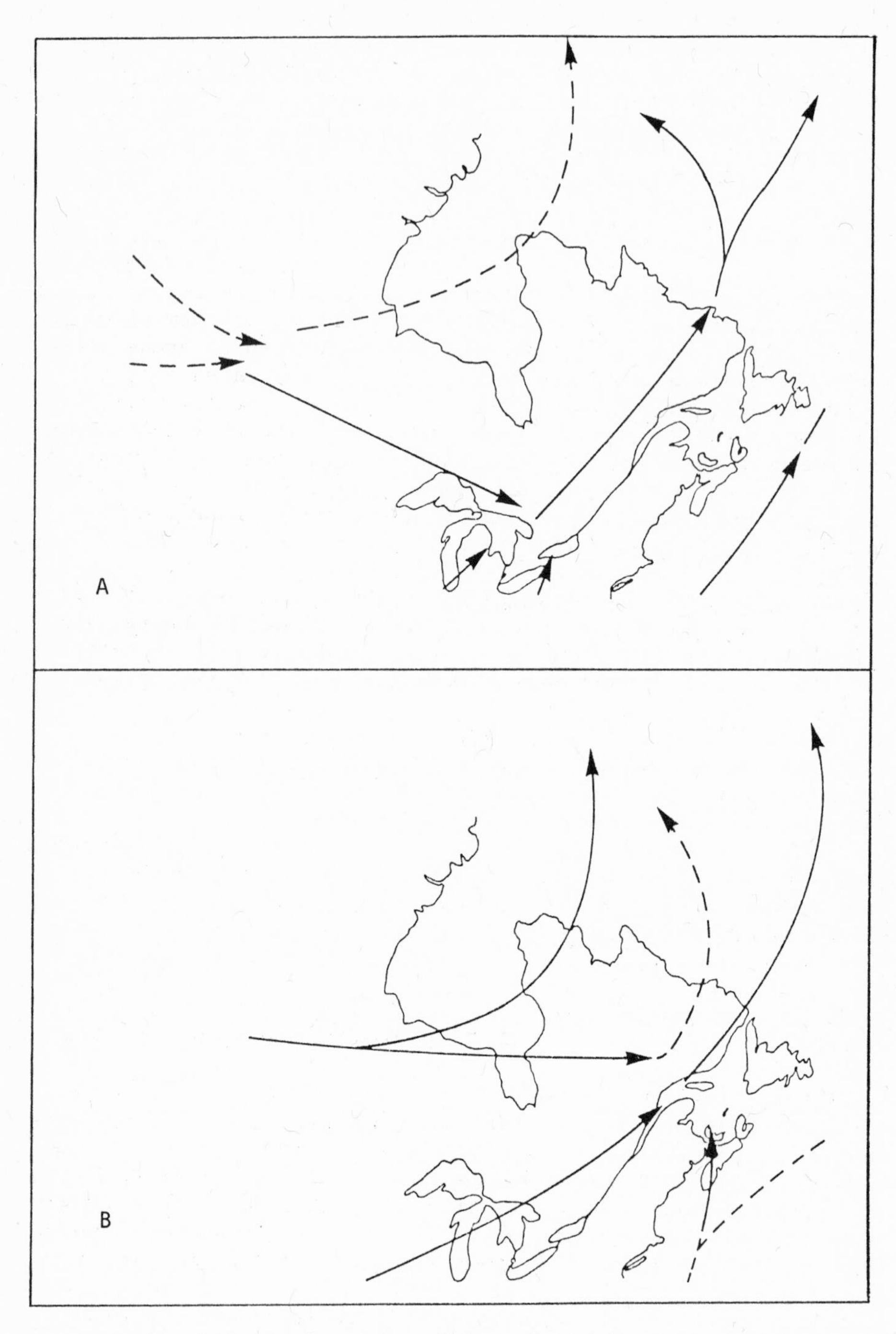

Fig. 3. Principal tracks of lows for (A) January-March and (B) July-September (from Archibald, 1956).

between the Pointe-des-Monts and the Rimouski sections. He found in the 1963 Pointe-des-Monts survey that there was a southerly transverse residual flow near the surface with velocities varying between 0.2 and 0.9 knots. This strong transverse current is the result of the existence of a cyclonic eddy west of Anticosti Island (El-Sabh, 1975, 1976). Along the north shore of the Gulf, surface water sets upstream, then follows the curve of the coast near Sept-Iles and swings across the estuary entrance towards the south shore. It then encounters the river outflow off that shore and, depending upon their relative strengths, either swings seaward to reinforce the Gaspé Current or turns upstream to create an eddy within the estuary. Farquharson found further evidence for the existence of such an eddy in the path followed over a 5-day period by a parachute drogue set to drift at a depth of 75 m.

The presence in the lower St. Lawrence estuary of a residual transverse flow of 0.14 knots toward the north shore in the upper layer, and of 0.02 knots in the same direction in the lower layer was first revealed by measurements of currents made by Forrester (1967, 1970) during 1965 in a cross-section near Rimouski. This northerly transverse current, together with the upstream currents in the middle of the section, strongly supports Farquharson's belief in the existence of an anti-cyclonic eddy within the estuary.

Using all the available current measurements, El-Sabh (1977) proposed a typical summer surface circulation pattern in the lower St. Lawrence estuary (Figure 4). The outflow of fresh-water along the south shore, the existence of an anti-cyclonic eddy with its centre somewhere between the Pointe-des-Monts and the Rimouski sections, together with the northerly transverse current in the Ile du Bic region are the main features of this circulation pattern. Waters from the Saguenay River set southward toward the south shore of the St. Lawrence estuary and join the fresh-water from the upper estuary which flows along that shore. This seaward flow along the south shore will continue until it reaches Ile du Bic region where it deviates toward the north shore with two branches: one which moves seaward along the north shore toward the Gulf, while the other turns to the west and moves upstream off Les Escoumins.

During the summer of 1975 and 1976, series of direct observations of currents were made at different locations in the estuary and under different meteorological conditions, in order to verify this circulation pattern and to obtain more insight and understanding on the mechanism involved, particularly in the area between Rimouski and Ile du Bic. Parachute drogues at 10 m depth were followed while their position and meteorological conditions were recorded every $\frac{1}{2}$ hour. In addition C.S.S. DAWSON was utilized to occupy a fixed station five miles off Rimouski for 4 days

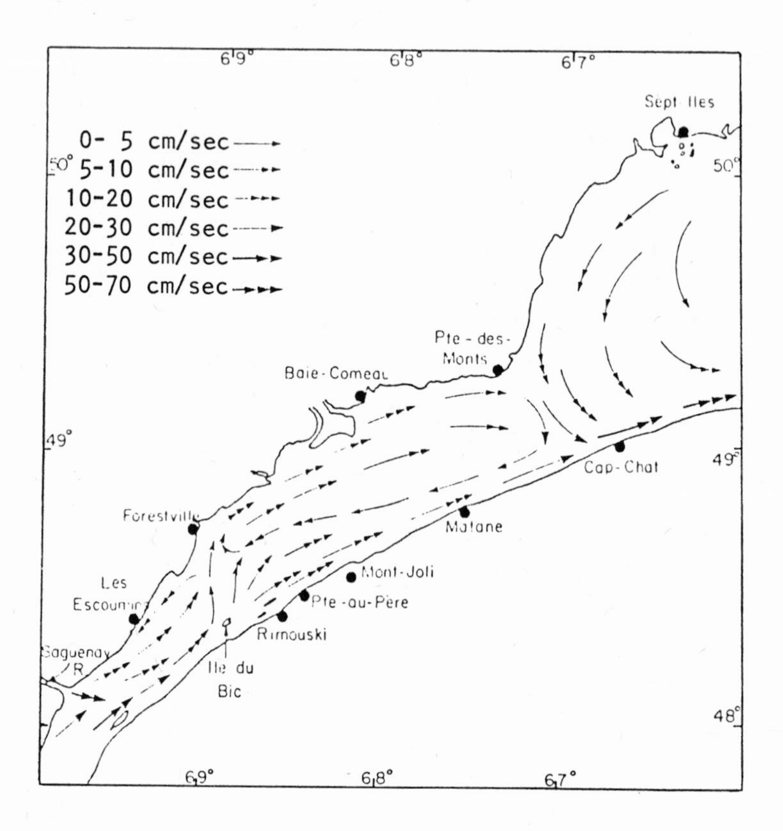

Fig. 4. Surface circulation pattern in the lower St. Lawrence estuary (modified from El-Sabh, 1977).

in August, 1975 during which temperature, salinity, currents and meteorological measurements were taken every 10 minutes. Preliminary analyses of these experiments indicate that winds play an important role in controling the water movements in the lower St. Lawrence estuary.

A parachute drogue placed near Ile du Bic in the south shore on October 18, 1975 and recovered after 32 hours near Forestville at the north shore (Figure 5) strongly confirms the existence of the northerly transverse residual current in that area with mean velocity of 0.6 knots. This experiment was carried out during calm conditions; the average wind velocity was 3 knots with a maximum of 8 knots, while wind direction varied between west and northeast (Figure 6).

The results obtained from the 4-day fixed station near Rimouski provide several points of interest and shed more light on the relation between the wind and current systems in the St. Lawrence estuary. Between 31 July and 4 August, 1975, two current meters of the Aanderaa type with velocity, conductivity, temperature and depths sensors, were moored at station 53, five miles off Rimouski, at depths of approximately 13 m below surface and 1 m above the bottom ($\sim$ 47 m). Intensive STD vertical profiles and meteorological conditions were taken every 15 minutes. Only the current and meteorological observations are presented here (Figures 7, 8 and 9). Looking closely at Figures 7 and 8, one can easily divide the direction of the residual currents recorded at 13 m depth into four periods which follow closely the meteorological conditions. Initially during calm weather (wind speed did not exceed 15 knots), the residual currents at 13 m were directed toward the north and northwest (transverse currents). During the second period, strong wind continued to blow from the west for almost 30 hours with average speed of 20 knots, while the direction of the residual current became southeast. Following a change in wind direction from west to east and northeast, with wind speed up to 30 knots, current velocities became higher and the residual currents were of greater magnitude than the tidal currents with direction toward the west. Finally, a change occurred in current direction to become southward following a change in wind direction which continued to blow from the north for 17 hours at the end of this experiment. It is of interest to notice that no correlation was found between variations in wind direction and currents observed at 47 m near the bottom (Figure 9), but rather a residual transverse flow of 0.11 knots toward the northwest.

From the above description of observed currents and wind, it is apparent that the wind can have an important influence on circulation pattern in the St. Lawrence estuary. With wind speed greater than 15 knots, the water current in the surface layer will

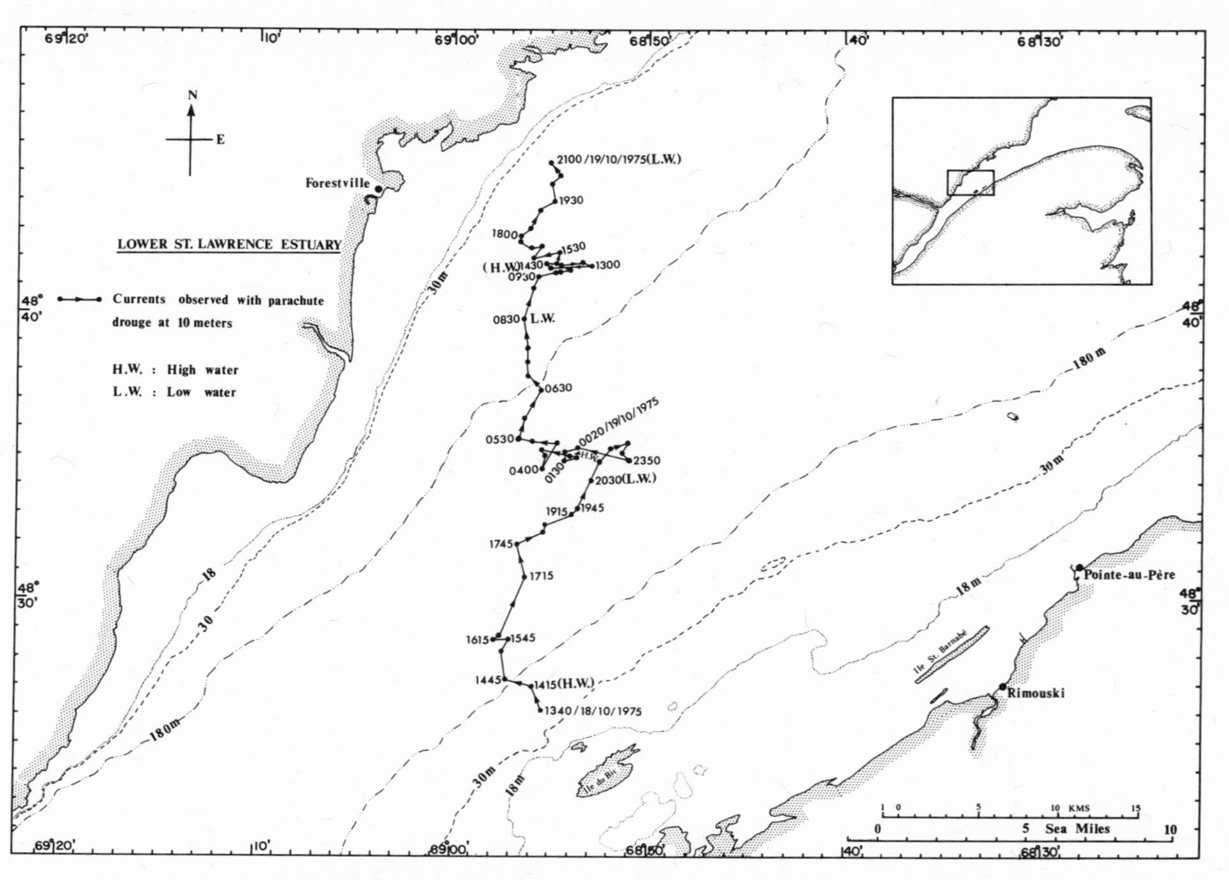

Fig. 5. Transverse currents in the lower St. Lawrence estuary observed with parachute drouge at 10 meters, during October, 1975.

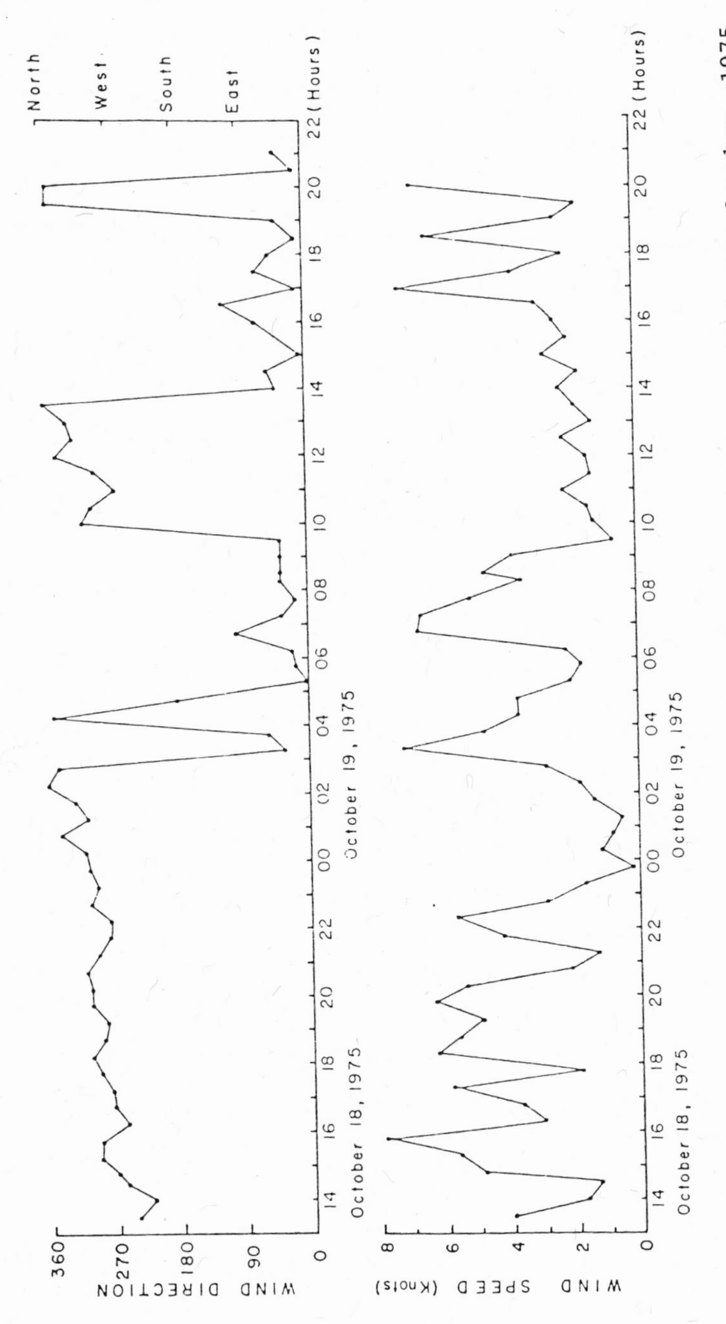

Fig. 6. Wind speed and direction in the lower St. Lawrence estuary during October, 1975.

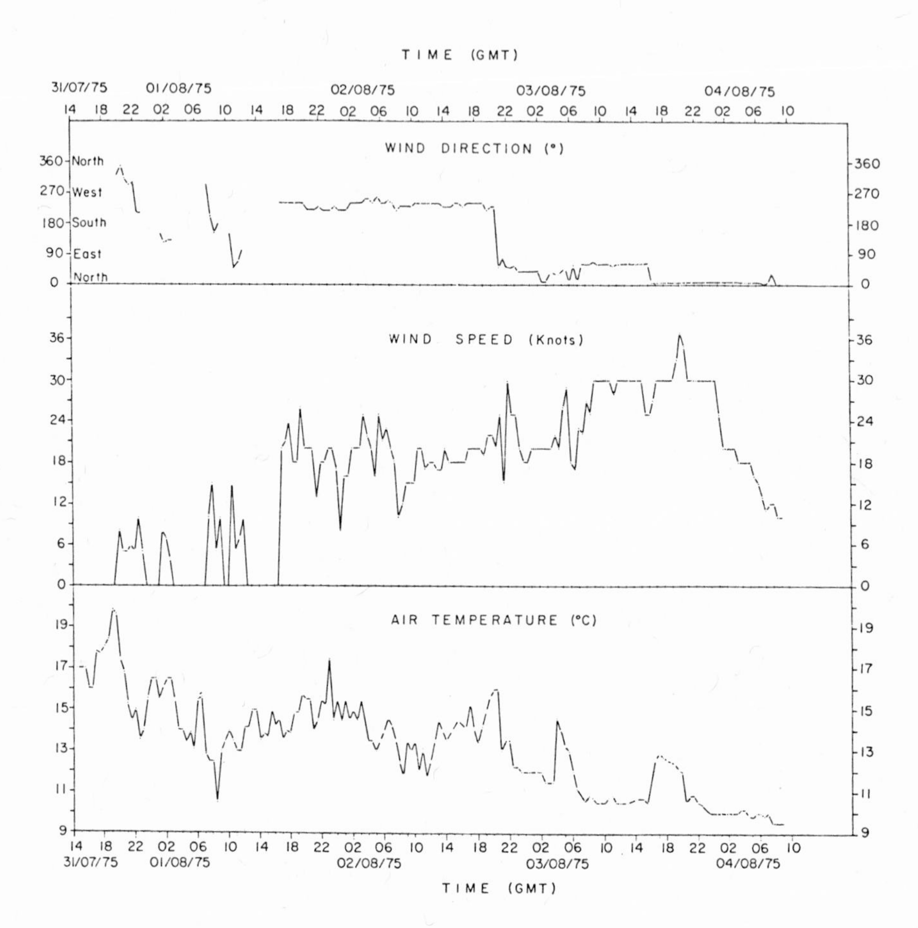

Fig. 7. Wind speed and direction and air temperature in the lower St. Lawrence estuary during August, 1975.

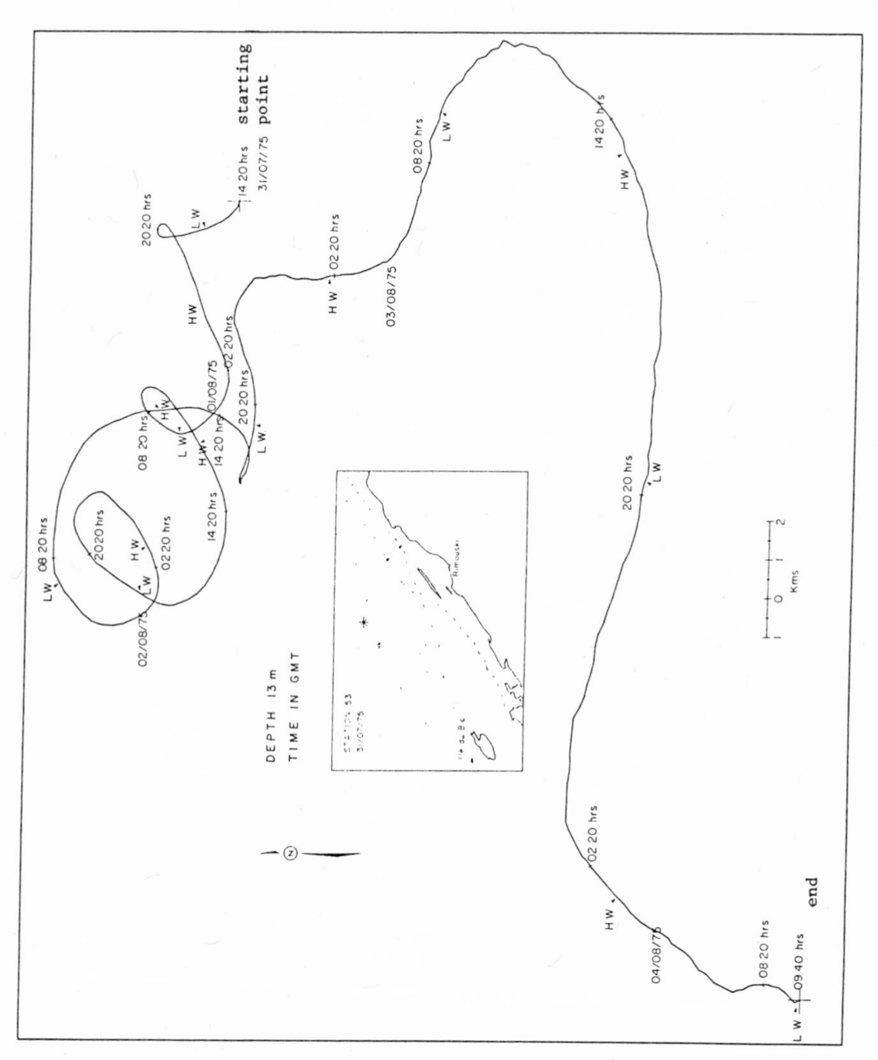

Fig. 8. Progressive vector diagram showing currents observed at station 53 near Rimouski at 13 m depth in August, 1975.

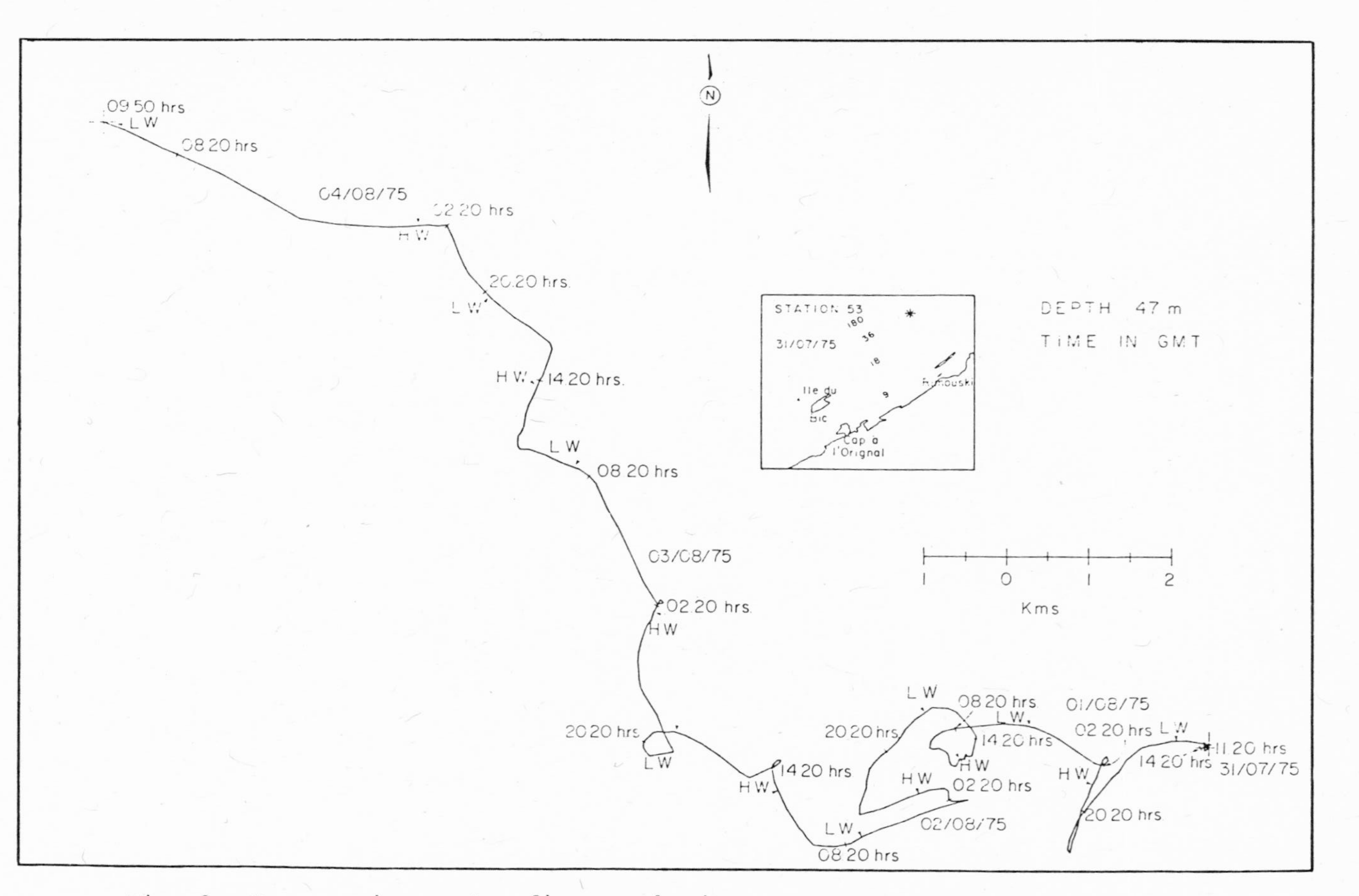

Fig. 9. Progressive vector diagram showing currents observed at station 53 near Rimouski at 47 m depth in August, 1975

be mainly in the direction of the wind. When the wind stops blowing, or during calm weather, the fields of pressure and velocity require redistributions and adjustment, and the transverse currents start to develop. Next, we will examine the concept of the classical Rossby Adjustment Problem to account for these currents.

ROSSBY ADJUSTMENT PROBLEM: THE HOMOGENEOUS CASE

Rossby (1938) studied the mutual adjustment of mass and velocity distributions in a current system which is being built up by a prescribed wind system acting upon a portion of the ocean surface. With reference to Figure 10, consider a homogeneous, incompressible ocean of uniform depth D_0 at rest. At time $t = 0$ winds impart a certain amount of momentum to a semi-infinite strip (i.e. the strip is of infinite length in the x-direction) of width 2a. The coordinate system is such that y is perpendicular to x and pointed to the left in the horizontal plane. The fluid between $y = +a$ and $-a$ has an average velocity u in the x-direction.

The momentum M per unit length of the current is $2\,a\,u_0\,D_0\,\rho$ where ρ is the density of water and u_0 is the initial current in the x-direction. Since no pressure gradient is available to balance the Coriolis force, the current will move to the right till a sufficient pressure gradient has been established to stop further movement to the right. Rossby studied only the final equilibrium state and not the transient aspects which were later considered by Cahn (1945).

During the process of adjustment, fluid columns to the left of the main current shrink in the vertical direction and stretch in the horizontal direction. Let y_0 and y respectively denote the initial and final positions of a fluid particle to the left of the main current. The equation of motion in the x-direction is:

$$\frac{du}{dt} = f\,v \tag{1}$$

where u and v are the velocity components in the x and y-directions and f is the Coriolis parameter. Note that friction is ignored and initially there is no pressure gradient. Integration with respect to t gives:

$$u = f\,(y - y_0) \tag{2}$$

Let D_0 and D respectively denote the initial and final depths of a fluid column. Then by continuity:

$$D\,.\,dy = D_0\,.\,dy_0 \tag{3}$$

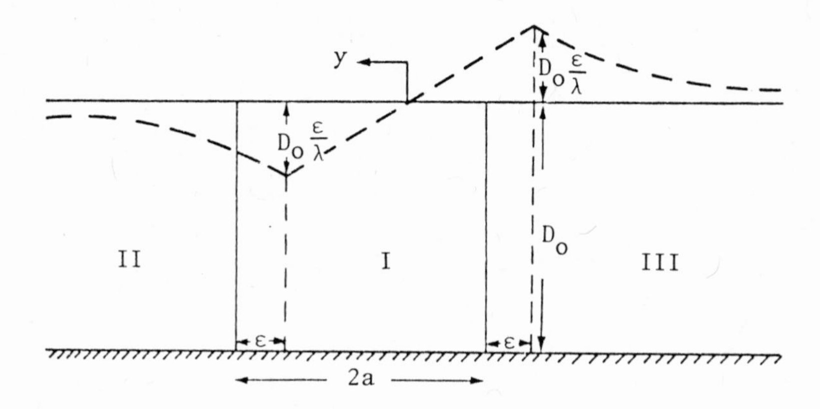

Fig. 10. Schematic representation of the adjustment processes for the homogeneous case (modifed from (Rossby, 1938).

After adjustment is completed, gradient motion is given by:

$$0 = - fu - g \frac{\partial D}{\partial y} \tag{4}$$

Define Rossby radius of deformation λ as:

$$\lambda \equiv \frac{\sqrt{g D_0}}{f} \tag{5}$$

Then from (2) to (5):

$$\frac{d^2 y_0}{dy^2} = \frac{y_0 - y}{\lambda^2} \tag{6}$$

A similar equation can be written for D (simply by replacing y by D in equation 6). This equation is a special case of the more general equation which expresses the conservation of absolute velocity:

$$\frac{\partial^2 D}{\partial x^2} + \frac{\partial^2 D}{\partial y^2} = \frac{D - D_0}{\lambda^2} \tag{7}$$

Following Rossby, integration of equation (6) with respect to y gives:

$$y_0 - y = A e^{y/\lambda} + B e^{-y/\lambda} \tag{8}$$

To determine the constants of integration A and B, invoke the condition that, the displacement $y_0 - y$ must vanish for large y. This gives A = 0 and if we denote by ε the total displacement to the right, of the left edge of the main stream, then:

$$y_0 - y = \varepsilon e^{-y/\lambda} \tag{9}$$

The heights D_ℓ and D_r of the free surface at the left and right edges of the current are given from equations (3) and (9) as follows:

$$D_\ell = D_0 \left(1 - \frac{\varepsilon}{\lambda}\right) \tag{10}$$

$$D_r = D_0 \left(1 + \frac{\varepsilon}{\lambda}\right) \tag{11}$$

Since the width of the current is 2a, the slope of the free surface across the current is $(D_r - D_\ell)/2a$. In the final equilibrium state, the mean velocity u_f of the main stream is:

$$u_f = \frac{g}{f}\frac{1}{2\,a}(D_r - D_\ell) = \frac{\lambda}{a} f\,\varepsilon \qquad (12)$$

The constancy of the absolute momentum of the main stream is expressed as:

$$\int_{-a+\varepsilon}^{a+\varepsilon} D_0\,(u_0 - fy)\,dy = \int_{-a}^{+a} D\,(u_f - fy)\,dy \qquad (13)$$

Noting that u_0, u_f, D_0 are constants. Equation (13) becomes after some algebra:

$$2\,a\,D_0\,(u_0 - u_f) = f\,D_0 \int_{-a+\varepsilon}^{a+\varepsilon} y\,dy - f\int_{-a}^{+a} Dy\,dy \qquad (14)$$

From equations (10) and (11) one can obtain:

$$D - D_0 = -\,D_0\,\frac{\varepsilon}{\lambda}\frac{y}{a} \qquad (15)$$

From equations (14) and (15) we have:

$$u_f = u_0 - f\,\varepsilon\,(1 + \frac{a}{3\,\lambda}) \qquad (16)$$

The total displacement to the right of the center of mass, during the adjustment process is $\varepsilon\,\{1 + (a/3\,\lambda)\}$. From equations (12) and (16):

$$\varepsilon = \frac{u_0}{f}\,\frac{1}{(1 + \frac{\lambda}{a} + \frac{a}{3\,\lambda})} \qquad (17)$$

ROSSBY ADJUSTMENT PROBLEM: THE STRATIFIED CASE

Next, following Rossby (1938) we will investigate the influence of stratification on the adjustment process. With reference to Figure 11, consider a two-layer ocean, with the upper and lower layers having initial thickness D_0 and D_0' and densities ρ and ρ'.

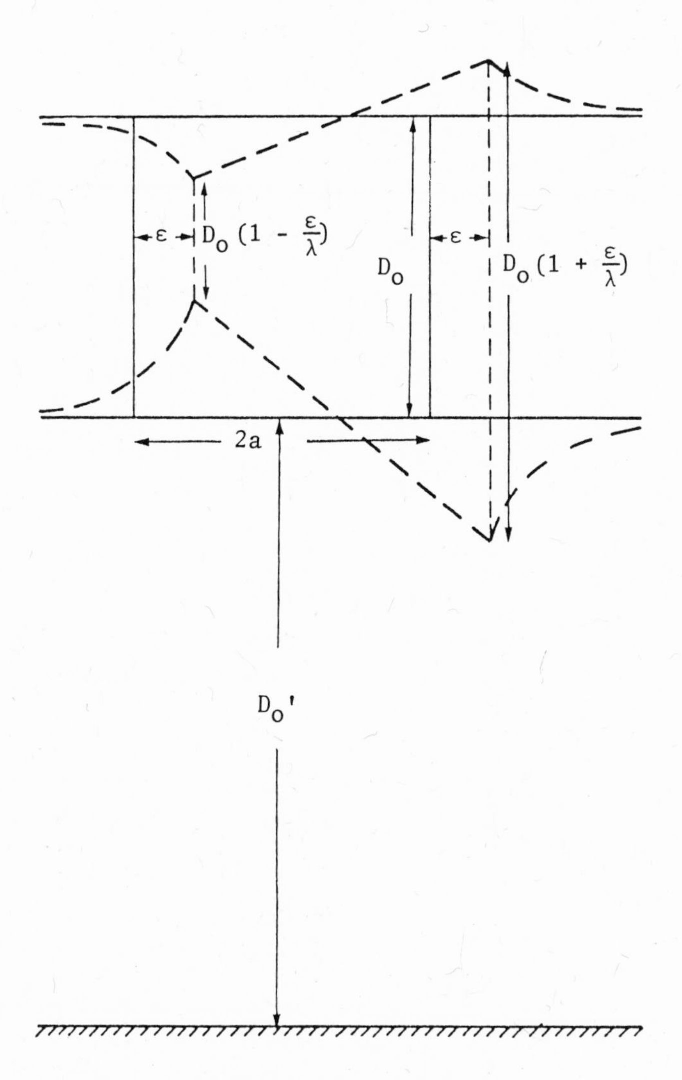

Fig. 11. Schematic representation of the adjustment processes for the stratified case (modified from Rossby, 1938).

It can be shown that the analysis in the previous section will also apply here, provided the so-called reduced gravity γ replaces the gravity g (Veronis, 1956), where:

$$\gamma = g \frac{(\rho' - \rho)}{\rho'} \tag{18}$$

The radius of deformation is now given by:

$$\lambda = \frac{1}{f} \sqrt{\gamma D_0} = \frac{1}{f} \sqrt{\frac{(\rho' - \rho)}{\rho'} g D_0} \tag{19}$$

The energy available for the inertia oscillations (in the homogeneous as well as the stratified cases) can be determined by noting that the initial energy of the system is given by:

$$E_0 = \rho a D_0 u_0^2 \tag{20}$$

and the final energy is given by:

$$E_f = E_0 \frac{(\lambda/a)}{(1 + \frac{\lambda}{a} + \frac{a}{3 \lambda})} \tag{21}$$

Thus the fraction of the initial energy that is available for inertia oscillations is:

$$E_{osc} = E_0 \frac{(1 + \frac{a}{3 \lambda})}{(1 + \frac{\lambda}{a} + \frac{a}{3 \lambda})} \tag{22}$$

Bolin (1953) examined the transient problem in a stratified incompressible ocean. He showed that certain combinations of density and vertical shear lead to a very prolonged adjustment process. Veronis and Stommel (1956) investigated the response of a two-layer ocean to a variable wind-stress and showed that the duration of the wind-stress has more or less determines the amount of energy absorbed by inertio-gravity oscillations. When momentum is added impulsively most of the energy goes into these oscillations whereas when it is added over several pendulum hours, a major portion of the energy goes into quasi-geostrophic currents.

Following Veronis (1956) we will formally develop the solutions for the two-layer problem. Assuming no lateral friction forces and no lateral frictional transfer of momentum across the interface the vertically averaged equations of motion and continuity in the top layer are:

$$\frac{\partial U_1}{\partial t} - f\,V_1 = \frac{\tau}{D_1} \tag{23}$$

$$\frac{\partial V_1}{\partial t} + f\,U_1 = -\,g\,\frac{\partial \eta_1}{\partial y} \tag{24}$$

$$\frac{\partial}{\partial t}(\eta_1 - \eta_2) + D_1\,\frac{\partial V_1}{\partial y} = 0 \tag{25}$$

where τ is the wind stress in the x-direction, D_1 is the depth of the top layer and η_1 and η_2 respectively are the deviations of the free surface and interface from their equilibrium positions (here D_1 and D_2 correspond to D_0 and D_0' in Figure 10). Define:

$$A \equiv \frac{\rho}{\rho'} \text{ and } B \equiv \frac{\rho' - \rho}{\rho'} = \frac{\Delta\rho}{\rho'} \tag{26}$$

The corresponding equations for the lower layer are:

$$\frac{\partial U_2}{\partial t} - f\,V_2 = 0 \tag{27}$$

$$\frac{\partial V_2}{\partial t} + f\,U_2 = -\,g\,[A\,\frac{\partial \eta_1}{\partial y} + B\,\frac{\partial \eta_2}{\partial y}] \tag{28}$$

$$\frac{\partial \eta_2}{\partial t} + D_2\,\frac{\partial V_2}{\partial y} = 0 \tag{29}$$

The wind stress was taken such that:

$$\tau = \begin{cases} \tau_c \text{ constant for } |y| < a \text{ and } 0 < t < t_0 \\ 0 \text{ for } |y| > a \quad \text{for } 0 < t < t_0 \\ 0 \text{ for all } y \text{ for } t > t_0 \end{cases} \tag{30}$$

The initial and boundary conditions are:

$$U_1 = V_1 = \eta_1 = U_2 = V_2 = \eta_2 = 0 \begin{cases} \text{for all } y \text{ at } t = 0 \\ \text{for all } t,\ y = \pm\infty \end{cases} \tag{31}$$

The system of equations given by (23) to (31) can be solved, for example by the method of normal modes (Charney, 1955). The equations governing the external mode (barotropic mode) are:

$$\frac{\partial U^e}{\partial t} - f\ V^e = \frac{\tau}{D} \tag{32}$$

$$\frac{\partial V^e}{\partial t} + f\ U^e = -\ g\ \frac{\partial \eta^e}{\partial y} \tag{33}$$

$$\frac{\partial \eta^e}{\partial t} + D\ \frac{\partial V^e}{\partial y} = 0 \tag{34}$$

where superscript e denotes external mode. Here:

$$D \equiv D_1 + D_2 \text{ and } \bar{\rho} = \tfrac{1}{2}\ (\rho + \rho') \tag{35}$$

The equations for the internal mode are formally similar to those of external mode if we replace g by the reduced gravity defined in (18).

The solutions for the two-layer case can be written, following Veronis (1956) as follows:

$$\begin{aligned} U_1 &= U^e + \frac{D_2}{D}\ U^i \\ U_2 &= U^e - \frac{D_1}{D}\ U^i \\ V_1 &= V^e + \frac{D_2}{D}\ V^i \\ V_2 &= V^e - \frac{D_1}{D}\ V^i \\ \eta_1 &= \eta^e + b\ \left(\frac{D_2}{D_1}\right)^2\ \eta^i \\ \eta_2 &= \frac{D_2}{D}\ (\eta^e - \eta^i) \end{aligned} \tag{36}$$

where superscript i denotes the internal mode. Veronis gave the solutions for the quasi-geostrophic case in terms of exponential functions and for the general case in terms of the Bessel functions.

ADJUSTMENT PROBLEM TREATMENT OF THE ST. LAWRENCE ESTUARY

We will start with an interpretation of the results for the homogeneous and stratified cases and apply them to the St. Lawrence estuary, particularly in the region of Rimouski (Figure 2). The following values are typical:

Latitude = 49° N

Average water depth = 18 meters

Half width of the current = a = 5 km.

Then from equation (5) the Rossby radius of deformation λ for the homogeneous case is about 122 km. Hence, for the homogeneous case $a/\lambda \sim 0.04$ and thus $<< 1$.

When one considers the stratification, the following values are typical:

$$\frac{\rho' - \rho}{\rho'} \sim \begin{cases} 0.011 & \text{for May} \\ 0.003 & \text{for September} \end{cases} \qquad (37)$$

Then from equations (18) and (19):

$$\lambda^* \sim \begin{cases} 12.8 \text{ km} & \text{for May} \\ 6.7 \text{ km} & \text{for September} \end{cases} \qquad (38)$$

where λ^* denotes the radius of deformation for the stratified case. Hence, for the stratified case:

$$\frac{a}{\lambda^*} \sim \begin{cases} 0.4 & \text{for May} \\ 0.8 & \text{for September} \end{cases}$$

i.e. unlike in the homogeneous case, a/λ^* is of the order of unity.

The solutions given by Rossby for the stratified case are valid only for the case $a/\lambda^* << 1$. Mihaljan (1963) gave asymptotic solutions which are valid for two cases:

$$\frac{a}{\lambda^*} << 1 \quad \text{and} \quad \frac{a}{\lambda^*} >> 1$$

we will return to this part of the discussion a little later.

Equation (17) which follows Rossby's treatment for the homogeneous case shows that for a current of one knot ($\sim$ 50 cm/sec), $\varepsilon \sim 5$ km. Based on the study by Veronis (1956), Table 1 is prepared to show the percentage of energy which goes into the quasi-geostrophic currents and the percentage that goes into the non-geostrophic inertio-gravitational oscillations (the transverse currents are a manifestation of these oscillations, as will be shown later).

Table 1. Percentage partition of energy between quasi-geostrophic currents (top triangle) and non-geostrophic inertio-gravitational oscillations (bottom triangle).

	Wind imparts momentum		
	Impulsively	Over 6 pendulum hours	Over 12 pendulum hours
Homogeneous case	95 / 5	99 / 1	99.9 / 0.1
Reduced gravity case	16 / 84	33 / 67	83 / 17
Two-layer case	26 / 74	49 / 51	89 / 11

The geostrophic part of the solution for the homogeneous case gives a strong current in the x-direction in the region $|y| < a$ and very weak and broad counter currents on either side. In the reduced gravity case, the current is weak in the center but is stronger near the edges and strong and narrow counter currents exist on either side. It can be shown (Vernois, 1956) that the cross-stream velocity (i.e. transverse currents) vanishes when the wind stops blowing for the geostrophic part of the solution. Hence, any transverse currents that exist after the passage of the weather system must be due to non-geostrophic effects (i.e. inertio-gravitational oscillations in the present case). Therefore, the observed transverse velocity is a representative parameter for these oscillations.

One of the interesting results is that the intense inertio-gravitational oscillations (and the transverse currents) need not be accompanied by large deviations of either the free surface or the interface. This is due to the fact that in the internal mode of behavior, the non-geostrophic motions are mainly inertial and do not require large redistributions of mass. It is precisely this property of the internal mode that can give rise to such large inertial oscillations. In the homogeneous case, the shift of the current in the transverse direction rapidly builds up pressure gradients which dissipates the transverse motion.

Through an asymptotic solution, Veronis (1956) showed that the transverse velocity which is inertio-gravitational in character has a period of half a pendulum day and its magnitude decreases with time as:

$$\frac{1}{\sqrt{ft}}\sqrt{\frac{(2t-t_0)}{(t-t_0)}}$$

where t is the time from the moment the wind stress started imparting momentum to the water and t_0 is the duration of the wind stress. All the studies on the Rossby Adjustment Problem have been made for the oceanic scale. For smaller and shallower water bodies such as the St. Lawrence estuary, the wind imparts momentum to shallower water columns and hence, the resulting motions will be stronger than in the oceans.

Mihaljan (1963) gave an exact solution to the Adjustment problem. With reference to Figure 10, he considered three different regions. Initially he assumed that u_0 (y) is uniform in region I and zero in regions II and III. In equation (6) he obtained an extra term $- u_0 (y)/(f \lambda^2)$ on the right hand side. Thus, Rossby ignored this term in his treatment. Mihaljan gave asymptotic solutions for the two cases $a/\lambda^* \gg 1$ and $\ll 1$.

We have shown above that for the St. Lawrence estuary a/λ^* is of order unity and no analytical solution is available for this case. We are developing a numerical model which includes the following features: depth variations, stratification, Coriolis force, friction, finite duration of the wind-stress, side boundaries and variation in the x-direction. The qualitative examination of the transverse currents in this estuary using the concept of the Rossby Adjustment Problem, attempted here points to the necessity of such a numerical model.

SUMMARY AND CONCLUSIONS

The concepts of the classical Rossby Adjustment Problem have been invoked to account for the transverse currents observed in the St. Lawrence estuary. The analytical solutions available till now are applicable for the oceanic scale and are relevant for the situations where the ratio of the half width of the current to the Rossby radius of deformation is either very much greater or very much less than unity. For the St. Lawrence estuary case, this ratio is of order unity and no analytical solution is available. Hence, the problem is examined here only qualitatively and a numerical model is being developed which will include boundries, depth variation, variable wind-stress, Coriolis force and friction.

ACKNOWLEDGEMENTS

This work was financially supported by the National Research Council of Canada, Grant A 3786 to M.I. El-Sabh and by a team grant from the Quebec Department of Education. The authors wish to thank Real Fournier for his help during the field operations and in preparing all the illustrations; Jocelyne Caron for the original typing and Margaret Johnstone for the final preparation and typing of the manuscript.

REFERENCES

Archibald, D.C. 1956. Intense storm tracks over Hudson Bay, the eastern Nova Scotia coast and the Grand Banks. Int. Rep. Meteorol. Br., Dep. Trans., Toronto: 8 pp.
Bolin, B. 1953. The adjustment of a non-balanced velocity field towards geostrophic equilibrium in a stratified field. Tellus, Vol. 5: 373-385.
Cahn, A. 1945. An investigation of the free oscillations of a simple current system. J. Meteorol., Vol. 2(2): 113-119.
Charney, J.C. 1955. The generation of oceanic currents by wind. J. Mar. Res., Vol. 14(4): 477-498.
El-Sabh, M.I. 1975. Transport and currents in the Gulf of St. Lawrence. Bedford Inst. Oceanogr., Rep. Ser. BI-R-75-9: 180 pp.
El-Sabh, M.I. 1976. Surface circulation pattern in the Gulf of St. Lawrence. J. Fish. Res. Board Can., Vol. 33(1): 124-138.
El-Sabh, M.I. 1977. Circulation pattern and water characteristics in the lower St. Lawrence estuary. In T.S. Murty [ed.] Proc. of Symposium on Modeling Transport Mechanisms in Oceans and Lakes, Environment Can. Rep. No. 43: 243-248.
Farquharson, W.I. 1966. St. Lawrence estuary current surveys. Bedford Inst. Oceanogr., Rep. No. 66-6: 84 pp.
Forrester, W.D. 1967. Currents and geostrophic currents in the St. Lawrence estuary. Bedford Inst. Oceanogr., Rep. No. 67-5: 175 pp.
Forrester, W.D. 1970. Geostrophic approximation in the St. Lawrence estuary. Tellus, Vol. 22(1): 53-65.
Forrester, W.D. 1974. Internal tides in the St. Lawrence estuary. J. Mar. Res., Vol. 32(1): 55-66.

Ingram, R.G. 1975. Influence of tidal-induced vertical mixing on primary productivity in the St. Lawrence estuary. Mem. Soc. R. Sci. Liege, 6e série, tome VII: 59-74.
Mihaljan, J.M. 1963. The exact solution of the Rossby Adjustment Problem. Tellus, Vol. 15(2): 150-154.
Rossby, C.G. 1938. On the mutual adjustment of pressure and velocity distributions in certain simple current systems, II. J. Mar. Res., Vol. 1(3): 239-263.
Veronis, G. 1956. Partition of energy between geostrophic and non-geostrophic oceanic motions. Deep-Sea Res., Vol. 3: 157-177.
Veronis, G. and H. Stommel. 1956. The action of variable wind stresses on a stratified ocean. J. Mar. Res., Vol. 15: 43-75.

APPENDIX: SOME PRELIMINARY NUMERICAL RESULTS

Veronis (1956) showed that the transverse velocity which is inertio-gravitational in character has a period of half-a-pendulum day. At Rimouski (Figure 1, latitude $\sim 49°N$) this is approximately 20 hours. With reference to Figure 8, the total duration of the direct current measurement was 91 hours 20 minutes (from 1420 hours on July 31, 1975 to 0940 hours on Aug. 4, 1975). An examination of Figure 8 shows that the currents have varied with a period of approximately 22 hours and 16 minutes which is quite close to the theoretical value of 20 hours.

In the homogeneous case (Cahn, 1945) the time taken for the establishment of the transverse currents is approximately equal to the time taken for a long gravity wave to travel the width of the estuary. If we take the width of the estuary at Rimouski as 60 km and the average depth as 18 meters, then the time taken to establish the transverse currents is about 1 hour 15 minutes. For the stratified case (September) using reduced gravity, the corresponding time is 22 hours 40 minutes which compares favorably with the observed value of 22 hours 16 minutes.

Some preliminary numerical calculations for the stratified case (September) with reduced gravity shows that the time taken for the establishment of the transverse currents is 22 hours 49 minutes which compares favorably with the observed value of 22 hours 16 minutes. Note that this period of approximately 22 hours corresponds to the lag between a change in wind direction (and speed) and the accompanying change in the current, whereas the much larger period of 7 to 10 days is the time taken for the total adjustment of the various fields starting from an initial state of rest.

ADVECTIVE TRANSPORT PROCESSES RELATED TO THE DESIGN OF WASTEWATER OUTFALLS FOR THE NEW JERSEY COAST

Joseph T. DeAlteris and Robert T. Keegan

Pandullo Quirk Associates

ABSTRACT

The results of selected portions of an oceanographic study are presented to elucidate the advective transport process related to the site selection and design of three ocean outfalls proposed for the southern New Jersey coast. Coastal currents which would transport a dispersing effluent plume, were measured using current meters, drogues, and drifters. Using a variety of analysis techniques, both spatial and temporal variations in the coastal current regime were investigated. Differences in the local rates of advective sediment transport result in erosion or accretion of the beach and sea floor. The long-term topographic stability was investigated using historical beach and offshore profile data. The short-term stability was studied using the results of monthly beach profiles, stake elevation measurements, and sediment sample analysis.

INTRODUCTION

Three ocean outfalls are proposed for Cape May County, New Jersey, as part of a Regional Wastewater Facilities Plan developed for the Cape May County Municipal Utilities Authority. Cape May County is located at the southern extremity of the State of New Jersey and the proposed outfalls will discharge into the coastal waters of the New York Bight area (Figure 1).

The Cape May County Atlantic Ocean Coastline is approximately 30 nautical miles long and consists of four elongated barrier islands, and a small peninsula of mainland at the southern end.

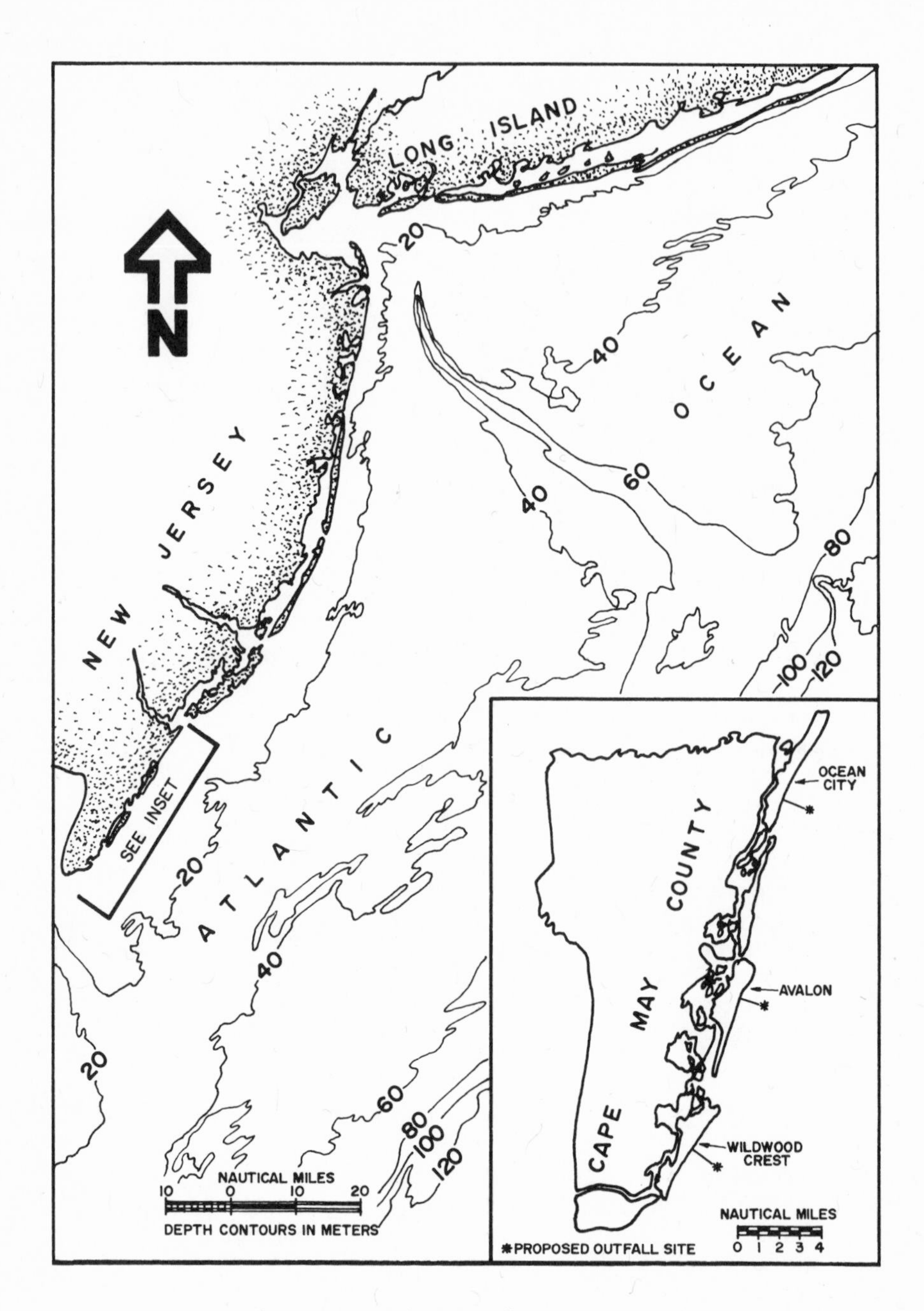

Figure 1. Regional Map of the New York Bight Showing the Cape May County Study Area

The coastal zone is populated and economically developed with summer tourist and commercial fishing industries. The ocean outfalls are proposed for the Peck Beach area of Ocean City in the northern portion of the region, Avalon in the middle of the region, and Wildwood Crest in the southern portion of the region, (see inset Figure 1).

A comprehensive physical oceanographic investigation was designed and implemented to measure the physical characteristics and the dynamics of the potential receiving body of water. A one-year field study included moored in situ recording current meters and thermographs, dye dispersion tests, wave, tide and meteorological measurements, drogue tracking exercise, surface and sea-bed drifter releases, and hydrographic surveys. A geological oceanographic analysis of beach and offshore sea-floor stability was conducted using the results of periodic beach and offshore profile measurements and a comparison of historical beach profiles and nearshore bathymetry. The results of this oceanographic study are being used in the final site selection and design of the outfall line and its diffuser and in assessing the potential impact of the outfall on the adjacent coastal environment.

Geostrophic and tidal currents, surface wind stresses, waves, and bottom frictional forces contribute to fluid motion in the coastal zone. This movement induces various transport processes that are of concern in the design of an ocean outfall (Figure 2). In the case of an outfall in which the discharged effluent has risen to the near-surface, the advective characteristics of the surface waters control the ultimate transport of the plume, which resulted from the near-field, jet-induced mixing of sewage effluent with ambient sea water. Motion of the sea near the bottom may also result in sediment movement. This transport phenomena can range from local non-advective, dispersive transport due to waves, to advective transport causing the migration of large scale bedforms, such as sand waves and linear ridges. Variations in the rate of sediment transport into and out of an area of sea floor result in elevation changes within that area. These small local changes when considered at a larger scale are reflected as erosion or accretion of the sea-floor. Observations of long and short term topographic changes in the sea-floor, from beach to offshore, are used to evaluate the historic and present stability of the sea-floor. The topographic stability of the sea-floor in the vicinity of the outfall alignment is an important parameter that must be considered in the design of a stable structure (Herbich, 1976).

The general ocean engineering considerations in the design of an ocean outfall are described by Beckman (1970) and Rainbow and Hennessy (1965). Many detailed investigations of the hydromechanical processes causing the reduction in concentration of the

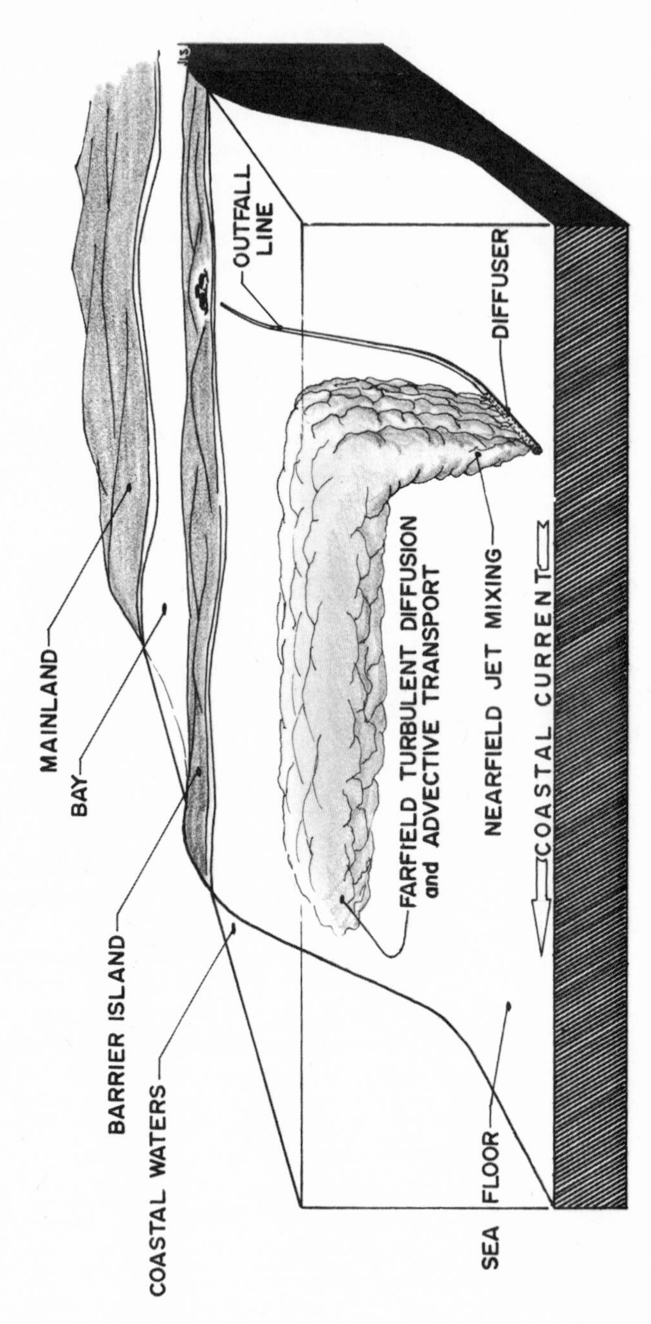

Figure 2. Treated Wastewater Effluent Plume in the Coastal Waters Having Been Discharged from the Outfall Line and Diffuser

sewage effluent are also described in the literature by Rawn and others (1960), Brooks (1960) and Sharp (1969). Studies of sediment transport and sea-floor stability are in the literature (DeAlteris and others (1975) and Madsen and Grant (1976)).

The purpose of this paper is to present the rationale, methodology, and results of selected portions of an oceanographic study designed to investigate the advective transport processes related to the final site selection and design of three ocean outfalls off the southern New Jersey coast. The intent is not to summarize the results of the study but to demonstrate the environmental engineering applicability of a study of this nature using specific examples from the Cape May County experience.

THE STUDY AREA

The Atlantic Coastal Waters offshore of Cape May County, New Jersey, are part of the southern portion of the New York Bight water mass.

The landward boundary of the Cape May Coastal Waters is approximately a 30 mile section of barrier island coast bounded on the north by Great Egg Harbor Inlet, bounded on the south by the entrance to Delaware Bay, and segmented by four other tidal inlets, Cold Spring, Hereford, Townsend, and Corson Inlets. The reversing tidal currents associated with these inlets and the irregular geometry of the barrier island coastline contribute to the complex dynamics of the coastal currents. The open seaward boundary of this area is the 60 foot (18 meter) depth contour located approximately 10 to 12 nautical miles offshore (Figure 1). There is a paucity of data on the currents along this boundary but an extrapolation of this data indicates this boundary to be characterized by weak rotary tidal currents of approximately 0.3 feet per second at a minimum, and a southwestward net drift current with a magnitude of less than 0.1 feet per second.

The sea floor slopes gently seaward from the shoreline to the 60 foot (18 meter) contour. The 30 foot (9 meter) contour lies approximately 1 to 2 nautical miles offshore. There are distinct linear sand ridges superimposed on the generally sandy bottom topography, especially nearshore, which may influence local water movement.

The U.S. Army Corps of Engineers (1972) describe tidal currents based on drogue studies 1 mile offshore of Wildwood Crest, New Jersey. Maximum velocities averaged 1 foot per second to the east-northeast during ebb tide in the Delaware Bay mouth. Studies by the Corps of Engineers (1972 and 1969) generally describe the historical changes in the beaches and offshore sea floor topography. In general, the shoreline is regressing landwards due to a rising sea level.

The coastal environment of Cape May County also lies within the Middle Atlantic Bight region. Bumpus and others (1973) have summarized the physical oceanography of this region; while Milliman (1973) has described the marine geology.

PROGRAM OF INVESTIGATION

Physical Oceanography

As stated previously, the "physical oceanographic" studies were principally concerned with the measurement and analysis of fluid motions in the coastal environment. These analyses were requried to establish dynamic patterns of the receiving waters, for prediction of the final destination and dilution of the effluent discharged from the outfall.

The coastal current regime is composed of unidirectional and oscillatory components. The oscillatory nature of the wave-generated currents do not result in a significant amount of net current, but storm-generated breaking waves would increase the dispersive characteristics of the water mass. The unidirectional currents are composed principally of tidal and wind-induced components. Along the coast of Cape May County, the tidal current characteristics vary considerably according to the location of adjacent tidal inlets and the distance from Delaware Bay mouth. Inlet influence is a maximum directly offshore of all the inlets, while the influence of Delaware Bay decreases with distance from the bay mouth.

The monitoring of currents was accomplished using both Eulerian and Lagrangian techniques (Figures 3 and 4). Because of the suspected wide spatial and temporal variations of coastal currents along the section of coastline, the use of both techniques was necessary. The current meter, a Eulerian device, records a time-series of current speed and direction at a single point. Drifters and drogues, Lagrangian devices, trace the movement of a water mass. Dye studies, another Lagrangian method, were used, not only to trace the water mass movement, but also to study the dispersive capability of the water.

An in situ instrument system was deployed at each of the three proposed ocean outfall locations for approximately a one-year period. The fixed instrument system provided a continuous time series of measurements specific to a point in space. These data were used to define temporal variations in water velocity and

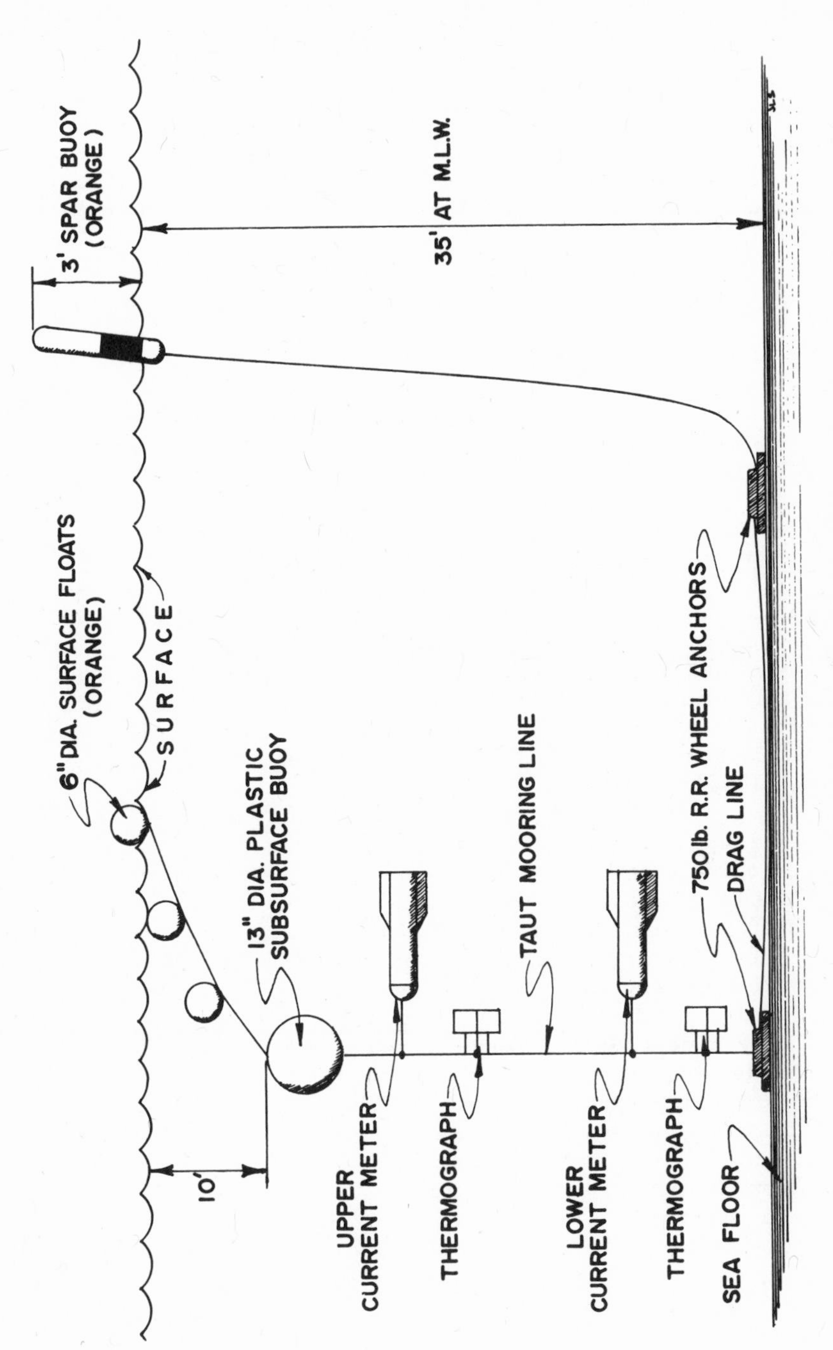

Figure 3. Eulerian Current Measuring System

water temperature at that site. Each array (Figure 3) consisted of a surface marker buoy and a taut-moor wire system. Attached to the taut-moor wire were two in situ recording ducted impeller current meters, upper and lower, and two in situ recording thermographs, upper and lower.

Drogue studies, using free floating current followers (Figure 4), were conducted at each of the three sites under consideration. These data were used to investigate spatial variability in surface currents in the vicinity of the in situ instrument stations. Currents associated with the adjacent tidal inlets may have significantly influenced both the magnitude and direction of the currents as measured at the in situ instrument stations. Therefore, the tracking of drogue paths during selected portions on the tidal cycle provided information on "water particle" trajectories (initial starting points being the in situ instrument stations). The advantages of drogues are that the mean velocity can be measured very accurately, as this is only a function of the start position, the end position, and time between observations, all precise knowns. The drogue is also a self-integrating device, that is, the effect of "small scale turbulence", superimposed on the mean flow, is eliminated. In these studies, the research vessel simultaneously tracked the trajectories or 4 to 5 drogues placed at depths ranging from 3 to 15 feet. Periodically, drogue position was determined using precision navigation equipment, and the data recorded as range-range values at a given time. This was converted to New Jersey plane coordinates, and velocity vectors determined and plotted. These were related to tide stage and geographic location.

Surface and sea-bed drifters, (Figure 4), were released on a random schedule at the three in situ instrument system locations using a small plane. At each drop site, six surface and six sea-bed drifters were released into the ocean. Each drifter was identified by a serial number so that when and if it was found, the point of recovery could be related to the point of release and the speed and direction of drifter travel determined. The surface drifters were made from an 8 ounce screw cap bottle ballasted with sand. Inside was placed a self-addressed, stamped post card for return to a local post office box. The surface drifter moves with the near surface currents, that is the upper 2 to 6 inches of the water column. It is important to note, that this region of the water column is most responsive to wind stresses, and therefore, the surface drifter is, in reality, a good indicator of surface wind drift currents. The sea-bed drifters were manufactured after the original Woodhead sea-bed drifter, an inexpensive flexible plastic disc with a weighted tail. It was constructed of a yellow polyethylene saucer shaped disc, 7 inches in diameter, with a bright red polyethylene tail 21 inches long attached to the center of the disc and weighted with a 5.5 gram special corrosion-

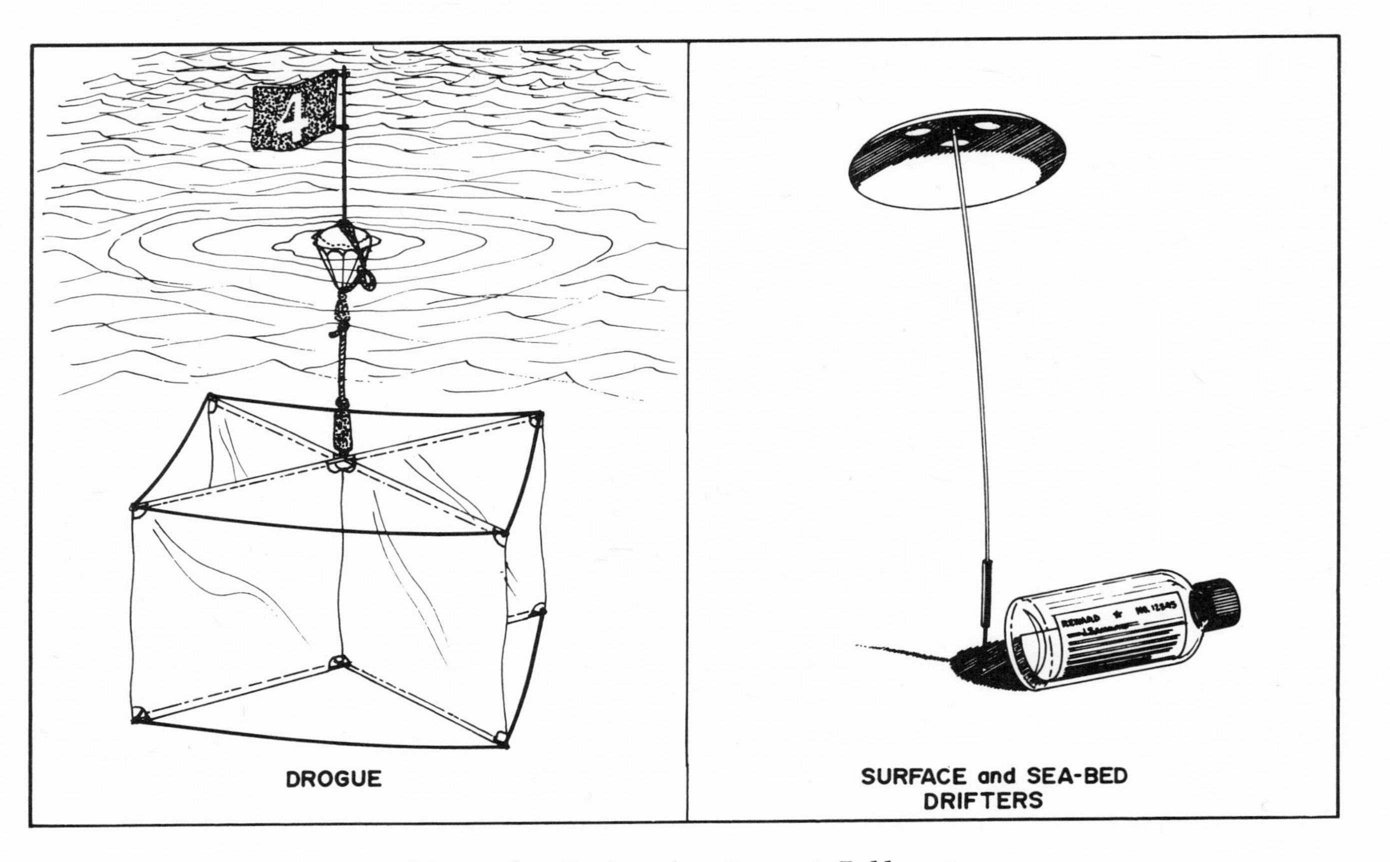

Figure 4. Lagrangian Current Followers

resistant tubing weight. The return legend and serial number were specially embossed onto the disc. This type of sea-bed drifter has been used successfully at Woods Hole Oceanographic Institute and elsewhere to measure bottom currents over extensive areas in continental shelf and estuary studies.

Geological Oceanography

The "geological oceanographic studies" were designed to assess the potential impact of the dynamic environment on the outfall lines and the diffuser structures. These studies included investigations into beach and offshore bottom stability. Three beach profiles in the vicinity of each of the proposed outfall alignments were measured monthly. The profile locations were specially selected in line with historical profiles dating to 1955 available from the U.S. Army Corps of Engineers. A comparison was also made between recent and historic bathymetric charts, along profiles extending seaward from the beach profiles. The measurement of stake fields by SCUBA divers (Figure 5) at the in situ instrument stations and the collection of sediment samples for sediment transport analysis were also conducted. Individual stake elevations provided a precise method of monitoring short-term sea floor topographic changes, while sediment textural data was used in assessing the dynamic processes.

RESULTS AND INTERPRETATIONS

Physical Oceanography

Data for the analysis of the coastal current regime was available from several sources, current meters, drogues, and drifters. Each source served a particular purpose, and when taken together provided a characterization of the coastal current regime as it would effect the advection of the effluent plume advection. For the purposes of the following discussion, selected results taken from the proposed Avalon site are discussed.

The current meter data was initially available as a time-series at one-half hour sampling intervals. The raw data was summarized into "Monthly Summary Histogram Plots," an example of which is shown in Figure 6 for August of 1975. The maximum observed current, as well as the calculated net current are illustrated. The results indicate that during this month and at this site, there was a small net current of 0.11 feet per second to the southwest compared to an average current speed of 0.4 feet per second. Ninety percent (90%) of the half hour average vectors were in the coast parallel direction indicating infrequent events

Figure 5. Scuba Diver Preparing to Measure a Stake Field

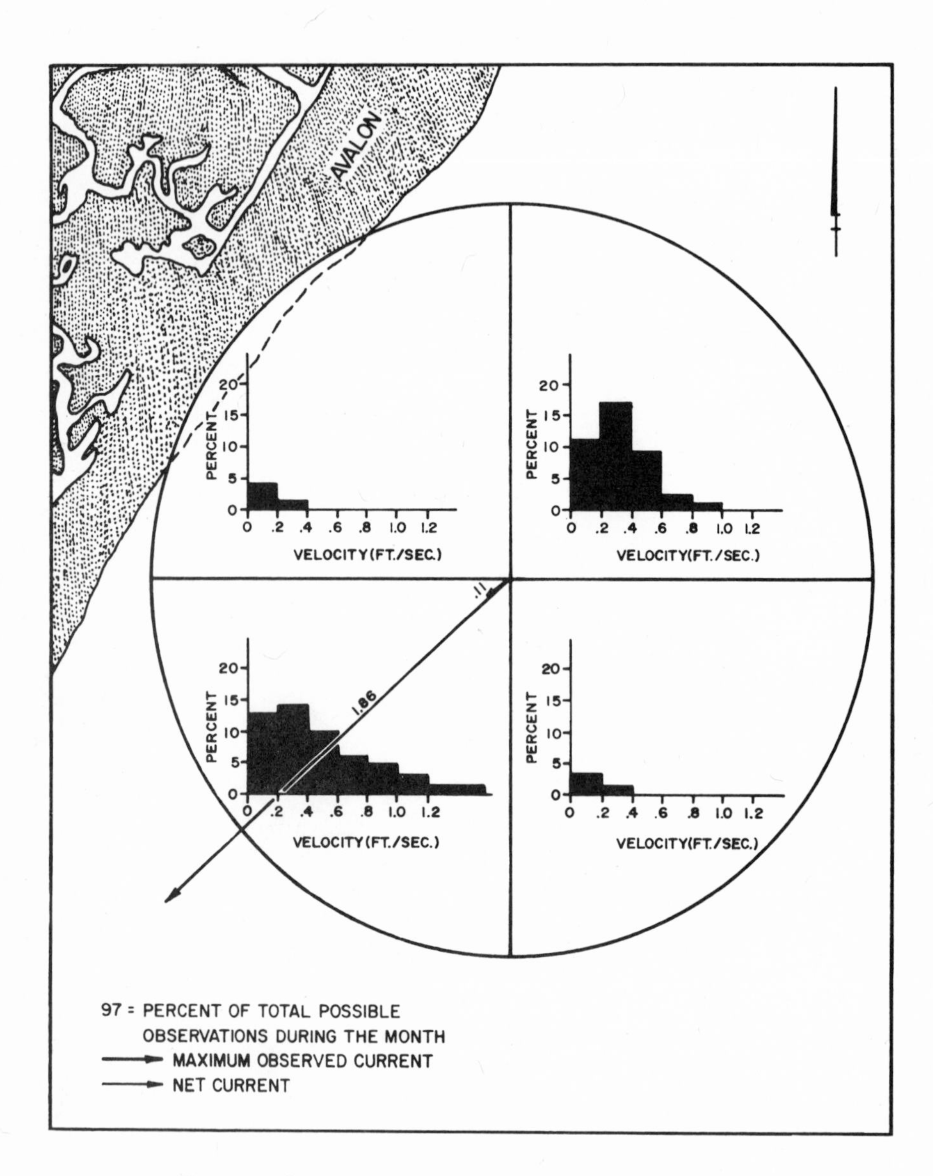

Figure 6. Monthly Summary Histogram Plot
Avalon Upper Current Meter, August, 1975

of direct onshore or offshore transport. This type of information is useful in the determination of the outfall diffuser orientation. In this case, a diffuser orientation perpendicular to the coast and the predominate coast-parallel currents, is suggested, in order to maximize the initial dilution of the effluent.

The single vector time-series was resolved into two separate scalar time-series, the on/offshore component, and the alongshore component. These series were then harmonically analyzed (Bloomfield, 1976) for the principal tidal constituents. In this study, six tidal constitutents with the indicated periods were used: S2, 12.00 hr.; M2 12.42 hr.; N2, 12.66 hr.; K1, 23.93 hr.; P1, 24.07 hr.; and O1, 25.82 hr. Using the resulting amplitudes and phase lags at the specified periods, the alongshore and on/offshore components of the tidal currents were predicted. Comparisons of predicted tidal and observed current components for upper current meter at the Avalon site suggested that the differences between the observed and predicted could be generally qualitatively correlated with local wind stress. A total tidal ellipse could be drawn on the locus of hourly current vectors during two tidal cycles. However, a more meaningful representation of the results is a progressive vector plot that shows the path or trajectory that a water particle would follow under the influence of tidal forces alone during two tidal cycles (Figure 7). From this plot, an estimate of the maximum distance the effluent plume will travel due to tidal forcing alone is obtained (assuming no spatial variation in the velocity field). In this case, the maximum tidal excursion was about 5,000 feet in the downcoast direction.

Other analyses of the current meter data included spectral or Fourier techniques and event analyses. The former technique was used to identify cyclic variations in the data including tides, and permited a correlation between observed cyclic variations and related forcing functions. In the Cape May data, a broad band peak with a 4 to 6 day period in the spectral density plots indicated the response of the coastal current regime to the passage of weather systems over the coast. The latter analysis method was used to describe particular events observed during the study, principally the effect of winds and barometric pressure on local tides and currents. DeAlteris and others (1977) report on observations during a severe northeaster and a hurricane, and contrast these to summer calms, periods of minimal mixing and advective transport.

Several drogue tracking exercises were conducted at each of the proposed outfall sites. These exercises ranged in duration from a half tidal cycle to two tidal cycles. The drogues were primarily used to investigate the spatial variability of the currents in the vicinity of the in situ instrument stations. This

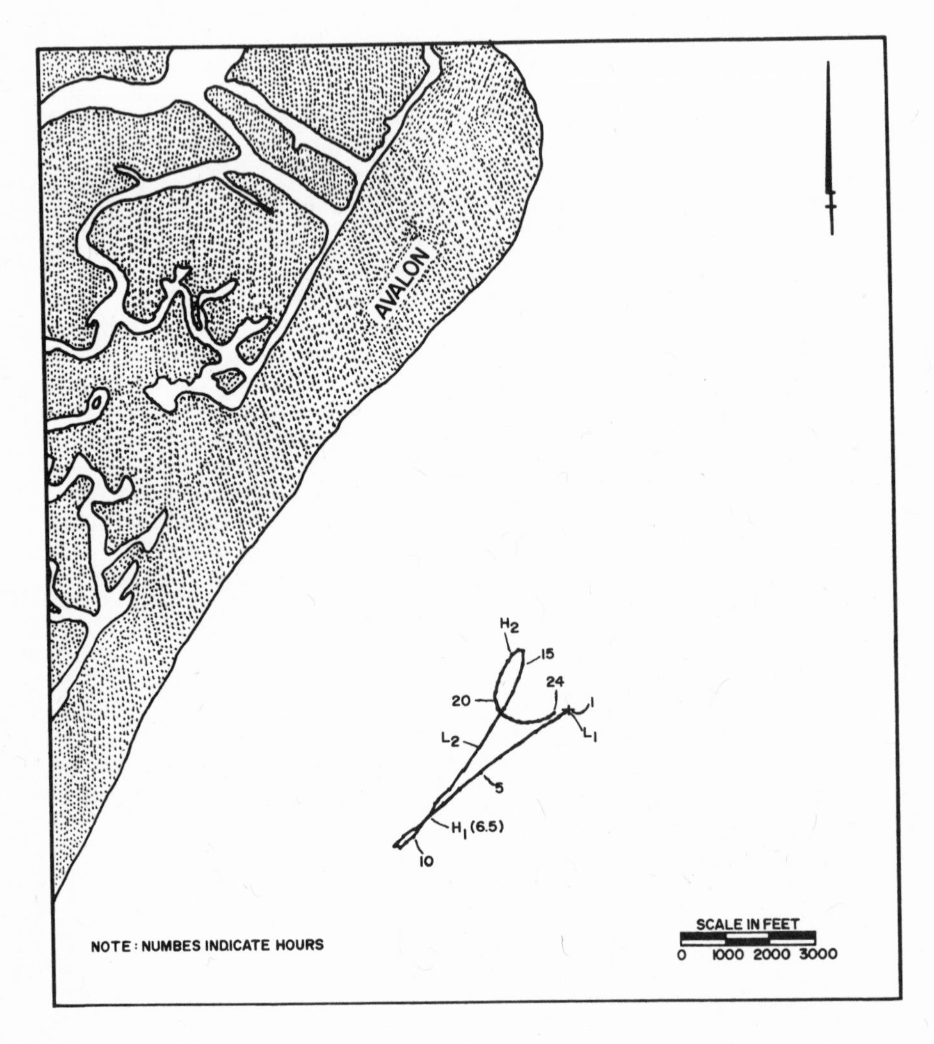

Figure 7. Progressive Vector Plot Predicted Tidal Currents Avalon Upper Current Meter, August-September, 1975

is especially important when the area is near coastal inlets. An inlet can have a major and possibly critical effect on currents which may not be reflected in the in situ data if the meter is placed outside of the range of the inlet's influence. For example, if an in situ current meter indicates predominately coast-parallel currents, but in reality strong on/offshore current due to an inlet exists a short distance away, the outfall's discharge, starting at the in situ instrument site could get "caught" by the inlet and be pulled towards the shore. Drogue studies would determine if the situations like this do indeed exist.

For each drogue release, four or five drogues were set adrift at depths ranging from 3 to 15 feet. The starting point of the drogue release was at one of the in situ instrument stations. An example of the results of the drift trajectories is shown in Figure 8. Also illustrated in the figure is the progressive vector plot of the upper in situ current meter data (recorded at the release point for the time during which the drogues were adrift). The tidal elevation plot as measured at Atlantic City was recorded during each drogue release and is also presented in the drogue trajectory figure. The purpose of the tidal elevation plot is to show the currents in relationship to the phase of the tide.

The results of the drogue releases were evaluated individually in light of the following information. For a short term observation, such as a drogue release, the most important influences on the currents are the tidal and wind driving forces. While a purely tidal current is nearly uniform from surface to bottom (ignoring bottom boundary layer effects); the wind-driven currents attain its maximum speed near the surface and diminishes with depth. A combination of these two types of currents can create various, complex vertical profiles.

Figure 8 presents results of the drogue study conducted October 1, 1975 at the proposed Avalon outfall site. Five drogues were released ranging in depth from 3 to 9 feet. The drogues were released at about low tide corresponding to maximum ebb current in Delaware Bay mouth. Winds were from the east at about 6 knots. All drogues and the progressive vector plot of the upper current meter behaved similarly. Initially, they moved in the upcoast direction. At approximately mid-tide, the drogues turned and moved in the downcoast direction for the remainder of the tracking period. All drogues as well as the progressive vector plot moved at a moderate speed, upcoast with ebb and downcoast with flood in the Delaware Bay mouth. Charlesworth (1968) first noted that possible influence of the Delaware Bay mouth on the tidal currents offshore of the southern New Jersey coast, "...as far as 20 miles north of Cape May Point...".

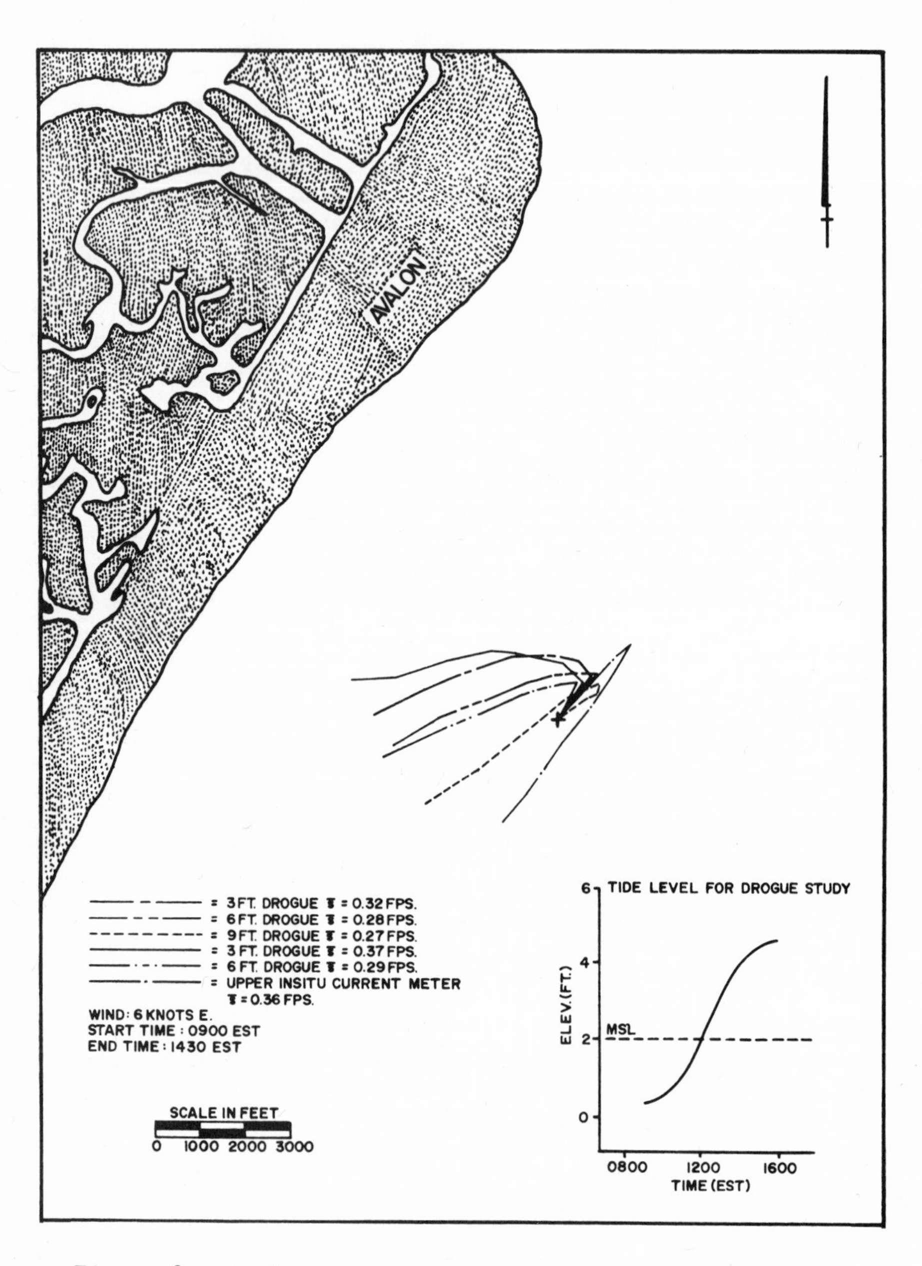

Figure 8. Avalon Drogue Study Results, October 1, 1975

The effects of the light east wind are clearly evident. The drogues at 3 feet have the greatest onshore drift components, while the drogues at 9 feet compare well with the progressive vector diagram from the upper current meter. The close comparison between the drogue at 9 feet and the upper current meter progressive vector plot indicates that in this case, there are no significant spatial variations in the horizontal velocity field and therefore data from the in situ current meter station is more useful for modeling purposes.

Of special importance in an oceanographic study for wastewater outfalls is the question: How often do the coastal currents have on onshore component that will transport a water parcel from the diffuser site 1.5 miles offshore to the beach area, and what is the time frame in which a water parcel will make that transit? The in situ recording current meters began to provide an answer to that question at least at the current meter station. The drogue tracking studies have further defined the area where a homogeneous flow field can be assumed to be reasonable. However, the existence and influence of the landward boundary can not be ignored, and there is no substitute for actual data on this transport phenomena from the offshore to the beach.

The surface and sea-bed drifters achieve this goal. Based on the percentage of drifters returned of the total released, the percentage of those returned indicating a direct onshore drift, and finally, the average number of days to recovery, one can obtain a "feel" for the percent occurrence and magnitude of coastal currents with a strong onshore component that will transport a drifter from the release site to the beach. Approximately 2800 drifters were released during the one year study period. Based on the returns of these drifters, the following statistics were calculated. The values of percent returned per time period of the total released at each site for the surface and sea-bed drifters are shown in Figure 9.

For the surface drifters, the percent returned ranged from a low of 16% for the Oct 1975 period at the Wildwood Crest release site, to a high of 90% for the Sep 1975 period at the Avalon release site. Although there appears to be no significant difference in returns between release sites, there definitely appears to be a time dependent trend in the return rate. During the Spring and Summer the average return rate is nearly twice that of the Fall-Winter return rate. During the entire study period, the surface drifters had an average return rate of approximately 46%.

For the sea-bed drifters, the percent returned ranged from a low of 11% for the Nov-Dec 1975 period at the Wildwood Crest release site to a high of 65% for Oct 1975 at the Ocean City release site. There appears neither significant differences in

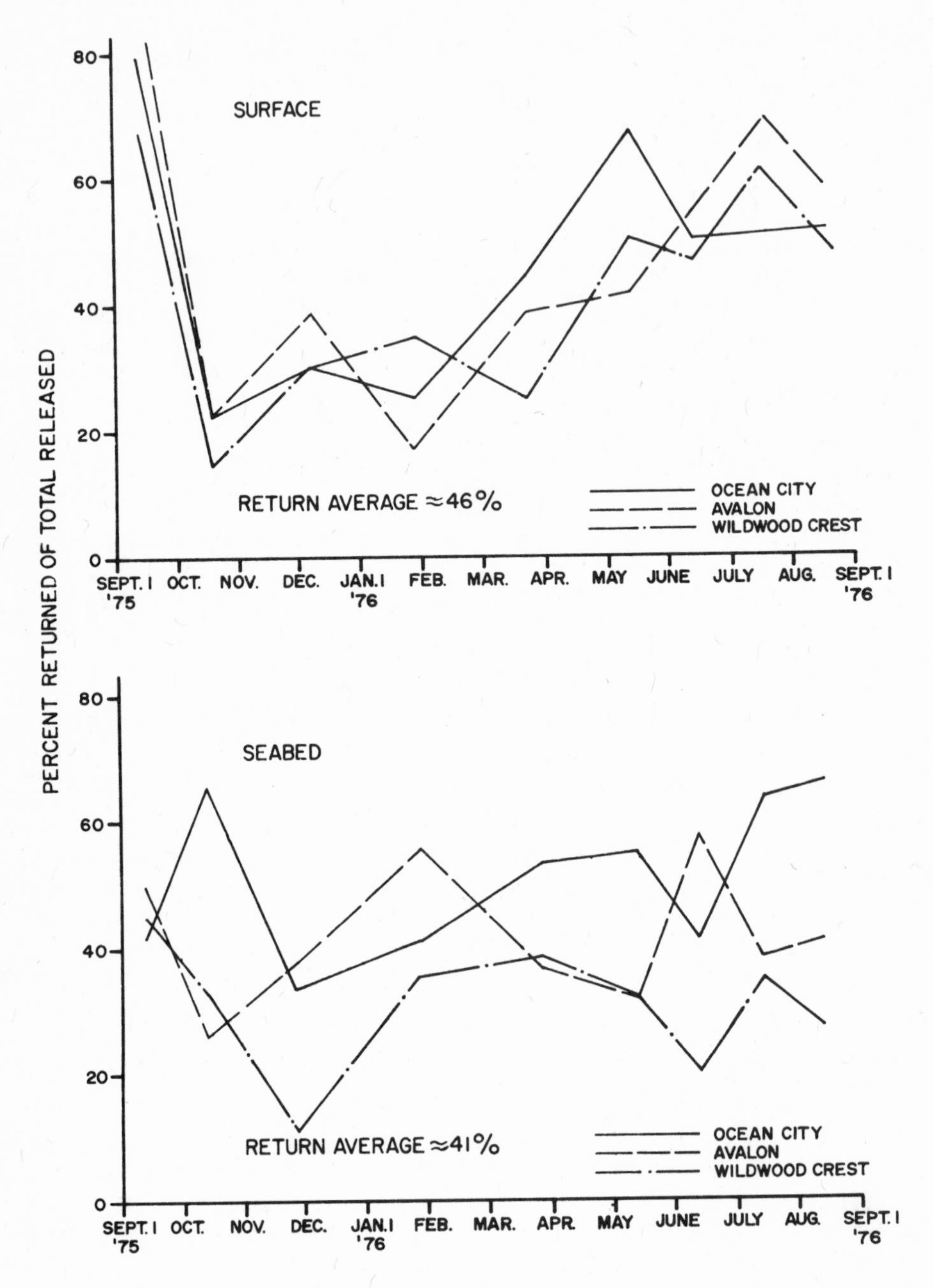

Figure 9. Surface and Seabed Drifter Returns

sites nor a time dependent trend in the percent returned values. The average sea-bed drifter percent return rate was approximately 41%, slightly lower than the return rate for the surface drifters. These return rates are significantly higher than others reported in the literature, for example, 25% (Schumacher and Korgen, 1976), 22% (Bumpus, 1965), and 18% (Harrison et al., 1967). This relatively high return rate is attributed to the short distance between the release site and the adjacent beach.

Individual drifters were recovered in locations ranging from Rhode Island, Long Island, Delaware Bay, to the Virginia coast. Although these returns are of interest, those drifters that were recovered directly onshore of their release site are of particular significance in this study. For the surface drifters, of those recovered, an average of 43% drifted directly onshore, with an average time of 2 days from release to recovery. For the sea-bed drifter, of those recovered, an average of 54% drifted directly onshore, with an average time of 7.5 days from release to recovery.

Assuming the time of recovery nearly coincides with the time of arrival of a drifter on the beach, then one can estimate the average drift velocity from the above statistics. For simplicity, we assume the release site is approximately 2 miles from the adjacent beach, then 20% of the surface drifters released (the product of 46% recovered and 43% of those with direct onshore drift) were recovered directly onshore of their release site with an average drift rate of 1 mile per day (less than 0.01 feet per second). For the sea-bed drifters, 22% of those released were recovered directly onshore of their release site with an average drift rate of 1/4 mile per day (much less than 0.01 feet per second). It is important to compare these drift rates to the observed currents in the Cape May coastal waters. Average maximum tidal currents are approximately 1 foot per second. The tidal currents range from 0.01 to 1.5 feet per second. Thus the velocity of the drifters from the site to the beach is far less than the observed currents at the sites. This result points out the importance of the boundary effect, resulting in a decreased onshore velocity, as water approaches the landward boundary.

Geological Oceanography

The geological oceanography studies were directed toward an assessment of the stability of the topography of the beach and seafloor along the proposed alignment of the outfall. Again, rather than attempt to describe the results of the study at each of the proposed sites, only a portion of the results of the study at the proposed Wildwood Crest site are discussed, with a presentation of the data for Profile No. 118, located approximately 1 mile south of the proposed alignment, (Figure 1).

Historically, the shoreline of Wildwood Crest has experienced a slight accretionary trend during the period of 1842-1975. This can be attributed to the artifical closing of Turtle Gut Inlet in 1917, and the subsequent construction of jetties at Cold Spring Inlet (U.S. Army Corps of Engineers, 1972). Note, this local trend is opposite to the overall trend of shoreline erosion experienced by the majority of the county.

The beach profile data used in the study included historical profiles from 1955, 1963, and 1965, available from the U.S. Army Corps of Engineers. The beach profiles were monitored monthly during the one year field study. For analysis purposes, the individual beach profiles were not as important as the envelope of observed change during the monitoring period. Thus, the short-term change (noise) could be compared to a long-term change (trend).

In the case of Profile No. 118, at Wildwood Crest, New Jersey, (Figure 10), the long-term trend (1955-1975) is one of stability, neither accretion nor erosion. Further, the short-term vertical fluctuations in elevation exceed those indicated by the long-term data. Thus, the conclusion is that in these case, short-term vertical changes in the profile on the order of 4 feet were observed and that this exceeded the long-term observed changes in the profile from 1955 to 1975.

Horizontal short-term changes in the profile were on the order of 100 feet at the mean low water mark (0′) and 200 feet at the mean high water mark (4′). As in the case of vertical change, the short-term envelope of observed horizontal change exceeded the indicated long-term horizontal change.

Reliable historical bathymetric data for Profile No. 118 from 1928, 1965 and 1976 were also available for analysis. The 1928 profile was taken from a U.S. Coast and Geodetic Survey Smooth Sheet. The 1965 profile was obtained from the U.S. Army Corps of Engineers. The 1976 profile was taken from field data collected during this oceanographic study using standard precision bathymetric survey procedures. A comparison of the profiles is shown in Figure 11. Observed vertical changes in the topography of the seafloor between the mean low water mark, approximately 500 feet from the profile origin, and the 35-foot depth, approximately 6500 feet from the profile origin, are quite small, on the order of 3 feet or less. Thus, historically, this portion of the profile appears relatively stable.

However, moving seaward from the 35-foot contour, the sea-

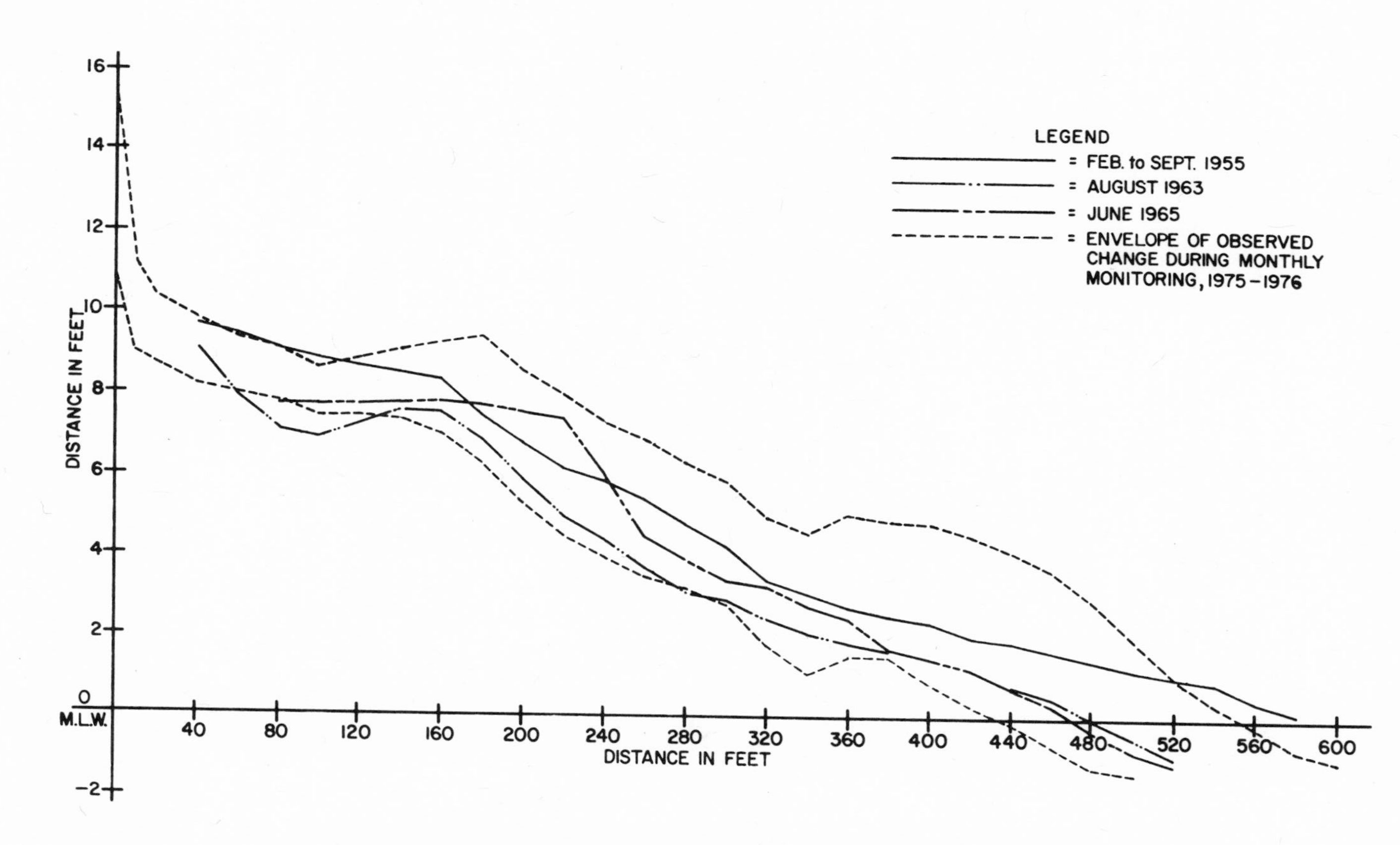

Figure 10. Historical and Short-Term Beach Profiles
Profile No. 118 - Wildwood Crest, New Jersey

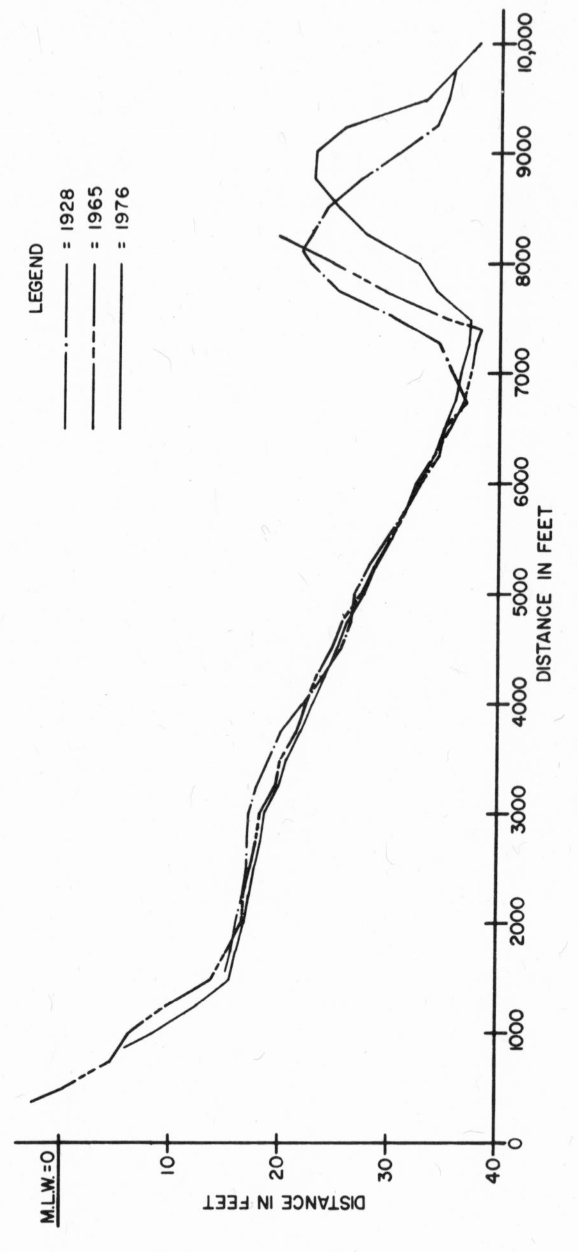

Figure 11. Historical Offshore Bathymetric Profiles
Profile No. 118 - Wildwood Crest, New Jersey

floor rises up over the crest of a linear ridge. The bathymetric profiles indicate that the ridge has moved seaward at a rate of approximately 25 feet per year. Concurrent with the seaward migration, the ridge crest has risen, and the landward trough between the ridge and the nearshore portion of the profile has filled slightly.

Fortunately, the proposed outfall alignment is north of this profile. There a stable profile extends to the 45-foot depth before rising over the ridge. However, the point to be garnered from this example is that although the nearshore portion of the profile appears quite stable, the seafloor is very dynamic even as far as 1-2 miles offshore. The mechanism for offshore linear ridge evolution and maintenance is a subject of debate in the scientific literature. Our purpose in investigating this area is not to attempt to answer a scientific question, but only to identify an area of potential concern to the outfall design engineer.

Having demonstrated that the offshore seafloor is dynamic, and either topographically stable or unstable, there remains the question of short-term stability. The stake fields and the sediment samples provided a unique source of information in this area. The results of the individual stake elevation measurements, for the proposed Wildwood Crest site, when plotted as a time series, (Figure 12), indicate very little change in the topography of the sea floor during the monitoring period. Textural analysis of the sediment samples taken by divers concurrently with the monthly stake measurements indicate that the cumulative grain size distribution of the surficial sediment plotted within a narrow envelope (not shown). The narrowness of the envelope between the curves is a measure of the temporal stability of the sedimentary environment at the measurement site. In this case, there appears to be a dynamic equilibrium between the hydraulic and sedimentary environments.

From the results presented in this section, several observations regarding advective sediment transport processes in the vicinity of a processed outfall can be made. The beach and offshore sea floor are dynamic areas. The results of the beach and offshore profiles, in this case, indicated long-term trends stability, neither accretion nor erosion, but a variability in vertical elevation of 2 to 4 feet. These observed changes in topography suggest a buried outfall line with the crown of the pipe and its armor below the maximum depth historically observed along the profile. Additional conservatism can be incorporated into the engineers design by specifying an even greater depth of burial, however this quickly reaches a point of diminishing return.

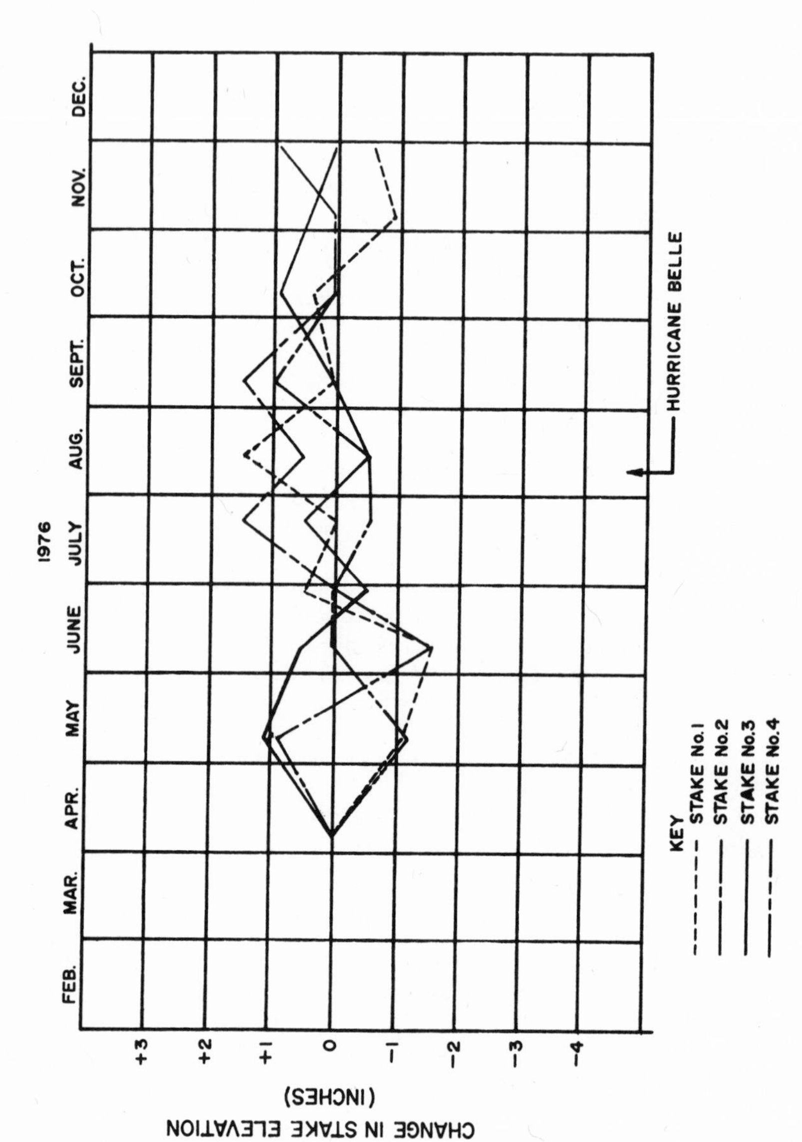

Figure 12. Short Term Sea Floor Topographic Changes Stake Field No. 3, Wildwood Crest, New Jersey

The design of an outfall can accommodate areas experiencing moderate shoreline erosion rates by again specifying a depth of burial that anticipates the projected erosion over the design life of the structure. In contrast, caution must be exercised in offshore areas, where large scale features such as linear ridges are migrating, especially in the vicinity of diffuser sections. Small changes in the topography of the sea floor could bury the diffuser ports, thereby reducing its usefulness.

SUMMARY AND CONCLUSIONS

The results of selected portions of an oceanographic study have been presented to describe the advective transport processes related to the final site selection and design of three ocean outfalls for the southern New Jersey coast. The intent has been neither to summarize the results of the total study nor to make a scientific contribution to a further understanding of the mechanics of the advective transport process themselves. The purpose of this contribution is to demonstrate the environmental engineering applicability of an oceanographic investigation of the advective transport processes.

Two physical phenomena have been considered: the advective transport of a dispersing effluent plume by coastal currents and the advective sediment transport on the beach and offshore sea floor, potentially resulting in topographic changes due to accretion or erosion of sediments.

Coastal currents were investigated for both spatial and temporal variations in the vicinity of the proposed diffuser. In situ recording current meters provided a continuous time series of data for analysis of temporal variations. Drogues were utilized to investigate local spatial variations; and drifters were used to study both spatial and temporal variations in adddition to long term drift patterns. The results of these studies have provided input to the final site selection of the outfalls, the required distance of the diffuser from shore, and the optimum orientation of the diffuser.

Rather than attempt to investigate actual advective sediment transport processes, the effects of sediment transport processes have been studied. That is, the stability of the topography of the beach and sea floor. Trends in the evolution of the beach and offshore sea-floor in the vicinity of a proposed outfall alignment were investigated using historical beach and offshore bathymetric profile data. Short-term variability in the topography of the beach and sea floor was studied using the results of a one year monthly beach profiling program and offshore stake elevation measurement program. The results of these studies have provided input to the final site selection of the outfalls, the depth of burial determination for the outfall, and the conceptual design of the diffuser structure.

REFERENCES

Abraham, G. and Brolsma, A.A., 1965, "Diffusers for Disposal of Sewage in Shallow Tidal Water": Pub. No. 37, Delft Hydralics Laboratory, 10p.

Beckman, W.J., 1970, "Engineering Considerations in the Design of an Ocean Outfall": Journal of Water Pollution Control Federation, Vol. 42, No. 10, pp. 1805-1831.

Bloomfield, P., 1976, "Fourier Analysis of Time Series: An Introduction": John Wiley and Sons, New York, 257 p.

Brooks, N.H. 1960, "Diffusion of Sewage Effluent in an Ocean Current": In, "Waste Disposal in the Marine Environment," Pearson, Editor, Pergamon Press, N.Y., 568 p.

Bumpus, D.F., 1965, "Residual Drift along the Bottom on the Continental Shelf in the Middle Atlantic Bight Area": Limnology and Oceanography Supplement, 10-R50-53.

Bumpus, D.F., and Lauzier, L.M., 1965, "Surface Circulation on the Continental Shelf": In, "Folio 7 of the Serial Atlas of the Marine Environment", American Geophysical Society, N.Y.

Bumpus, D.F., Lynde, R.E., Shaw, D.M., 1973, "Physical Oceanography": In, "Coastal and Offshore Environmental Inventory, Cape Hatteras to Nantucket Shoals", Marine Pub. #3, University of Rhode Island, R.I., 72 p.

Charlesworth, L.J., 1968, "Bay, Inlet, and Nearshore Marine Sedimentation; Beach Haven and Little Egg Inlets, New Jersey": Ph.D. Dissertation, University of Michigan, 247 p.

DeAlteris, J.T., Roney, J.R., Stahl, L.E., and Carr, C., 1975, "A Sediment Transport Study, Offshore, New Jersey": Proceedings Ocean Engineering III, American Society of Civil Engineers, N.Y., pp. 225-244.

DeAlteris, J.T., Shahrabani, D., Keegan, R., Kroll, J., 1977, "Currents of the Coastal Boundary Waters Off Southern New Jersey": Proceedings of Hydraulics of the Coastal Zone, American Society of Civil Engineers, New York, (in press).

Harison, W., Norcross, J., Pore, W. and Stanley, E.M., 1969, "Circulation of Shelf Waters off the Chesapeake Bight: Surface and Bottom Drift of Continental Shelf Waters Between Cape Henelopen, Delaware, and Cape Hatteras, North Carolina, June 1973-December 1974": ESSA Prof. Paper No. 3, pp. 1-81.

Herbich, J.B., 1976, "Scour Around Model Pipelines Due to Wave Action": Proceedings of 15th Conference on Coastal Engineering, American Society of Civil Engineers, New York, (in press).

Madsen, O.S. and Grant, W.D., 1967, "Sediment Transport in the Coastal Environment": Report No. 209, Department of Civil Engineering, M.I.T., 105 p.

Milliman, J.D., 1973, "Marine Geology": In, "Coastal and Offshore Environmental Inventory, Cape Hatteras to Nantucket Shoals", Marine Pub. #3, University of Rhode Island, R.I. 91 p.

Rainbow, C.A. and Hennessy, 1965, "Oceanographic Studies for a Small Wastewater Outfall": Journal Water Pollution Control Federation, Vol. 39. No. 11, pp. 1471-1480.

Rawn, A.M., Bowerman, F.R., and Brooks, N.H., 1960, "Diffusers for Disposal of Sewage in Seawater": Journal of Sanitary Engineering Division, Proceedings American Society of Civil Engineers, Vol. 86, SA2, pp. 65-105.

Schumacher, J.D. and Korgen, B.J., 1976, "A Sea-Bed Drifter Study of Near-Bottom Circulation in North Carolina Shelf Waters": Estuarine and Coastal Marine Science, Vol. 4, pp. 207-214.

Sharp, J.J., 1969, "Spread of Buoyant Jets at the Free Surface": Journal of Hydraulics Division, Proceedings American Society of Civil Enginers, Vol. 95, HY3, pp. 811-825.

U.S. Army Corps of Engineers, 1972, "Study of New Jersey Coastal Inlets and Beaches, Interim Report on Hereford Inlet to the Delaware Bay Entrance of the Cape May Canal": Philadelphia, Pennsylvania.

U.S. Army Corps of Engineers, 1969, "Study of New Jersey Coastal Inlets and Beaches, Great Egg Harbor Inlet to Stone Harbor": Philadelphia, Pennsylvania

PHOSPHORUS TRANSPORT IN LAKE ERIE

Ralph R. Rumer, Jr.

University of Delaware

ABSTRACT

The role of phosphorus as a causal factor in the eutrophication of Lake Erie is reviewed and the phosphorus budget analyzed. Recent studies by others provide quantitative estimates of the phosphorus loading to Lake Erie. In 1975 the total phosphorus loading to the lake was estimated to be 35,307 metric tons—3 percent from the atmosphere, 43 percent from shoreline erosion, 7 percent from the Lake Huron outflow, 24 percent from point sources within the Lake Erie drainage basin, and 25 percent from diffuse sources in the Lake Erie drainage basin. The above estimates are assumed to include phosphorus in all of its forms, i.e. total phosphorus. The disposition of total phosphorus in Lake Erie includes a mean value of 879 mg/kg in the sediments (based on dry sediment weight) and a mean value of 0.026 g/m^3 in the water column (although there were significant variations with position in the lake).

Mathematical representations for phosphorus budgets are reviewed and an adaptation of an earlier model by Lorenzen et al. (1976) is described in detail. The model incorporates sedimentation of particulate phosphorus to the sediments, release of phosphorus from the sediments, and retention of phosphorus in the sediments. Phosphorus uptake by phytoplankton is not specifically included. Specific rate constants are estimated for a three-basin version of Lake Erie based on available field data and assuming equilibrium conditions existed in 1975.

The concept of phosphorus retention in lakes is discussed and various analytical expressions for computing phosphorus retention are reviewed. It is shown that significant errors may result in computing phosphorus retention if equilibrium is assumed when not applicable. Phosphorus loading limits are discussed in relation to achieving a desired trophic status in Lake Erie. The findings strongly indicate that diffuse source loading of total phosphorus will need to be reduced if the desired improvements in the trophic status of Lake Erie are to be achieved. Alternative phosphorus loading reduction plans are presented to illustrate the application of the trophic status approach towards rehabilitating Lake Erie. Much research and data collection will be needed to more accurately quantify the phosphorus budget of Lake Erie and predict the response of the Lake Erie ecosystem to changes in phosphorus loading.

One important avenue of research is the coupling of the time-dependent hydrodynamic transport equations with the time-dependent material balance equations including the biochemical transformations resulting from phytoplankton growth. Present limitations to the development of such models include inadequacy of field data for model calibration purposes, limitations in computational capability, and insufficient knowledge regarding the analytical representation of the biochemical transformation and growth kinetics for phytoplankton.

INTRODUCTION

A significant problem area in Lake Erie deals with the high level of biological productivity and associated seasonal occurrences of oxygen depletion in some parts of the lake (Burns and Ross, 1972). Taste and odor problems in public drinking water supplies taken from the lake are also associated with this biological productivity. This high level of biological productivity is stimulated by nutrient enrichment of the lake water resulting from man's activities in the drainage basin, with phosphorus being the nutrient of primary concern. Hutchinson (1957) has discussed the phosphorus cycle in lakes and stated that "...(phosphorus) is in many ways the element most important to the ecologist, since it is most likely to be deficient...." More recently, Vollenweider (1971) stated that "...Phosphorus is usually the initiating factor (for eutrophication)...." Paul Sager (1976) in a recent review of lake eutrophication also underlined the relative importance of phosphorus. In this chapter, the phosphorus budget of Lake Erie is reviewed along with methodologies that could be used to evaluate the in-lake effect of reducing phosphorus loading to the lake.

PHOSPHORUS LOADINGS TO LAKE ERIE

Phosphorus enters the surface of the lake from the atmosphere, at the lake boundaries from shoreline erosion of phosphorus-bearing sediments, and in the tributary inflows that drain the watershed. The potential export of phosphorus into the lake from sources in the drainage basin is great. In fact, an ongoing study has concluded that management of diffuse sources of phosphorus (as well as point sources) will be required in order to achieve the desired rehabilitation of Lake Erie (U. S. Army Corps of Engineers, 1975).

Estimates of the atmospheric loading of phosphorus to Lake Erie range from 615 metric tons per year (Chapra, 1976) to 900 metric tons per year (U. S. Army Corps of Engineers, 1975). Both of these estimates were based on extrapolation of measurements made over other lakes. Burns et al. (1976a) report an estimate of phosphorus input to Lake Erie from shoreline erosion to be 14,800 metric tons per year. This estimate is based on the measurement of the phosphorus content of shoreline materials and of the quantity of shoreline materials eroded. As a result of an intensive sampling program of selected tributaries to Lake Erie, the U. S. Army Corps of Engineers (1975) has estimated the tributary loading to Lake Erie to be as follows: 2,334 metric tons per year from the Lake Huron outflow and 17,273 metric tons per year from the Lake Erie drainage basin itself. This later estimate includes 8,562 metric tons per year from point sources and 8,711 metric tons per year from diffuse sources. These estimates may be considered as representative of the phosphorus loading to Lake Erie in 1975. They are summarized in Table I.

Table I

Summary of Estimated Phosphorus* Loadings to Lake Erie, 1975
(see text for source of estimates)

Atmospheric Loading	900 metric tons
Shoreline Erosion	14,800 metric tons
Lake Huron Outflow	2,334 metric tons
Point Source Loading in Lake Erie Basin	8,562 metric tons
Diffuse Source Loading in Lake Erie Basin	8,711 metric tons
Total Loading	35,307 metric tons

*Expressed as total phosphorus.

The above estimates of phosphorus loading to Lake Erie are presumed to include phosphorus in all of its forms. There is doubt as to what proportion of this total phosphorus would be available for biological uptake. Burns (1976b) suggests that the phosphorus in eroded shoreline materials is not available because of its inertness. The U. S. Army Corps of Engineers (1975) state that shoreline eroded phosphorus is bound as apatite and not available for entry into the phosphorus cycle of Lake Erie. Future studies will shed more light on this question of availability; however, for the present, the phosphorus contributed from shoreline erosion is excluded from the analysis of biological productivity in Lake Erie. The contribution of total phosphorus from atmospheric loading is also excluded since it represents less than 5 percent of the remaining load.

Lake Erie has been frequently referred to as being composed of three sub-basins on the basis of its morphology. The three sub-basins are depicted in Figure 1, along with their physical measurements. The remaining loadings to Lake Erie are summarized in Table II according to sub-basin and whether point or diffuse (U. S. Army Corps of Engineers, 1975).

Table II

Remaining Total Phosphorus Loading to Lake Erie
(in metric tons per year)

Basin	Lake Huron Outflow	Point	Diffuse	Total
(1)	2,334	5,861	5,086	13,281
(2)	--	2,081	2,439	4,520
(3)	00	620	1,186	1,806
Totals	2,334	8,562	8,711	19,607

There is still a question as to what proportion of the remaining total phosphorus loadings to Lake Erie (given in Table II) is available for biological uptake. Because of this uncertainty, detailed budgets of the various forms (soluble reacting, total dissolved, particulate and apatite) of phosphorus are difficult to formulate. One recourse is to do a material balance on total phosphorus only.

DISPOSITION OF PHOSPHORUS IN LAKE ERIE

Williams et al. (1976a) have categorized the forms of phosphorus found in the surficial sediments of Lake Erie as apatite phosphorus, nonapatite inorganic phosphorus, and organic phosphorus. The distribution of sediment phosphorus within the lake did not correlate with locations of major phosphorus inputs. Rather, the distribution was governed by sediment properties such as particle size whose distribution, in turn, is governed by the hydrodynamic behavior of the lake. The mean value for sediment total phosphorus was found to be 879 mg/kg with a range of 188 mg/kg to 2,863 mg/kg (based on dry sediment). In general, regeneration of phosphorus from the sediments into the water column is not significant under oxic conditions (Williams et al., 1976b). However, under anoxic conditions the regeneration of phosphorus is significant, having been found to be eleven times as great as under oxic conditions

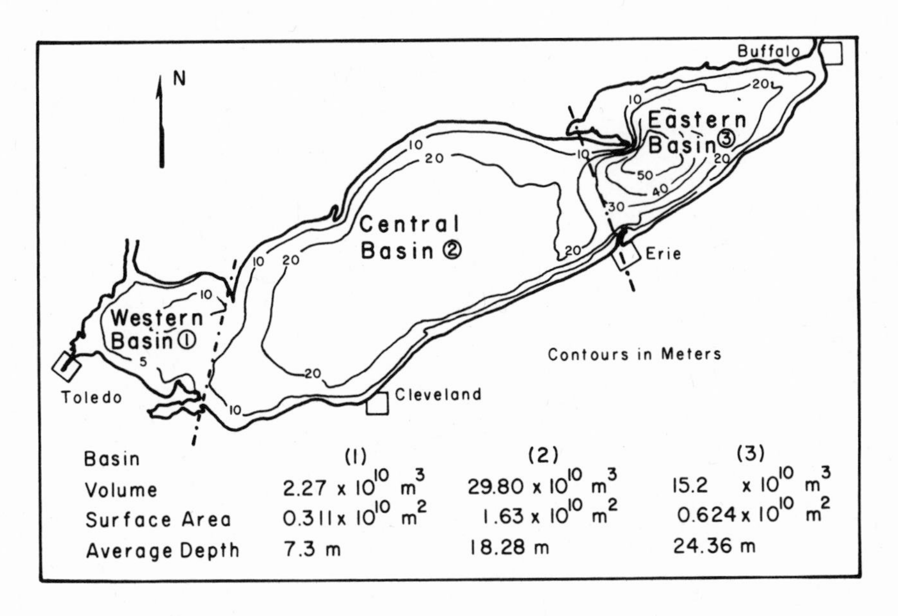

Basin	(1)	(2)	(3)
Volume	2.27×10^{10} m^3	29.80×10^{10} m^3	15.2×10^{10} m^3
Surface Area	0.311×10^{10} m^2	1.63×10^{10} m^2	0.624×10^{10} m^2
Average Depth	7.3 m	18.28 m	24.36 m

Figure 1. Lake Erie Bathymetry and Sub-Basins.

(Burns and Ross, 1972).

Particulate phosphorus constituted 50 to 70 percent of the total phosphorus found in the water column. The western basin averaged 66 percent, the central basin averaged 61 percent, and the eastern basin 56 percent (Burns, 1976c). The quantity of soluble reactive phosphorus varied widely throughout the lake with the western basin and the south shore of the central and eastern basins exhibiting the highest concentrations. The hypolimnetic waters of the central basin were observed to develop increases of soluble reactive phosphorus during anoxia, a direct result of phosphorus regeneration from the sediments. It was also observed that severe wind storms generated sufficiently high bottom currents in certain parts of the lake such that sediment was physically resuspended. Regeneration of phosphorus accompanied this storm activity, although subsequent sedimentation removed most of that regenerated.

The phosphorus incorporated in the lake biomass varies daily and seasonally. Estimates of this quantity of phosphorus depend upon estimates of the total biomass in the lake as well as knowledge of the percentage of the biomass composed of phosphorus. Vollenweider (1971) cited figures for the mean phosphorus content of phytoplankton ranging from 0.19 percent to 0.5 percent of the dry weight of the phytoplankton biomass. Using the reported molecular formula for phytoplankton cell material (i.e. $C_{106}H_{263}O_{110}N_{16}P$), the computed percentage is 0.87 percent, which is somewhat higher (Richards, et al., 1965). Reported basin wide concentrations of biomass in Lake Erie average about 4 g/m^3 (wet weight) (Munarvar and Munarvar, 1976). Assuming the dry weight to be 25 percent of the wet weight, the dry weight concentration of biomass would be on the order of 1 g/m^3 giving a biomass phosphorus concentration of 0.0087 g/m^3. Based on these very rough estimates, the phosphorus in the biomass could be on the order of 50 percent of the total phosphorus measured under certain conditions. This is based on reported average concentrations for total phosphorus as given in Table III (U. S. Army Corps of Engineers, 1975).

Table III

Estimated Total Phosphorus Concentrations in Lake Erie, 1975*

Western Basin	0.037 g/m^3
Central Basin	0.018 g/m^3
Eastern Basin	0.022 g/m^3
Whole Basin	0.026 g/m^3

*see U. S. Army Corps of Engineers, 1975

The disposition of phosphorus is portrayed schematically in Figure 2. The phosphorus incorporated in the biomass is considered as part of the total phosphorus in the lake water. The exchanges are indicated by arrows.

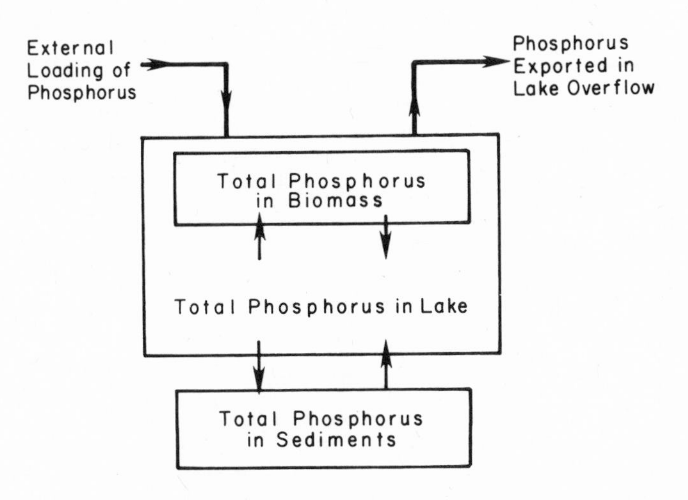

Figure 2. Schematic Representation of Phosphorus in Lake Erie.

This portrayal of the phosphorus budget of Lake Erie forms the basis for consideration of a methodology that might be used for estimating the in-lake effects (in terms of changes in concentrations of total phosphorus) that would result from specific reductions of phosphorus input to the lake; i.e., a management tool that could be used to evaluate the efficacy of various phosphorus load reduction schemes that might be implemented in the drainage basin.

REVIEW OF PHOSPHORUS MODELS

The objective of controlling phosphorus loading to Lake Erie is not, of course, only to reduce in-lake total phosphorus concentrations below some prescribed value. Rather, the objective is to reduce the high levels of biological productivity in the lake and, thereby, restore the lake to a more desirable trophic status. If the level of biological productivity were to be addressed specifically as the in-lake parameter to be predicted, then a modeling effort would be required that incorporated phytoplankton growth, death, and decay. Such models are being developed (DiToro et al., 1973) and show considerable promise both as frameworks for achieving a better understanding of lake ecosystems and, ultimately, as management tools. Such models are designed to simulate the lake as portrayed in Figure 2, including the biomass dynamics. In order to adequately represent the phytoplankton dynamics in these models, other nutrients must be accounted for as well as predation by zooplankton. The effects of water temperature, pH, and available sunlight must also be included. Readers are referred to the works of Thomann et al. (1974) for more information on the development of phytoplankton models.

Hornberger et al. (1975) evaluated a phytoplankton mathematical model by conducting nutrient enrichment experiments in a reservoir near Charlottesville, Virginia. They concluded that the principal application of phytoplankton models is for qualitative determinations and cautioned that each such model must be designed for the specific waterbody under consideration. They found the determination of "correct" growth parameters to be the most difficult aspect of model verification. Their calibrated model was used to predict the effects of nutrient enrichment with poor success.

O'Melia (1972) has discussed phosphorus modeling and pointed out the importance of sedimentation and the vertical exchange between the hypolimnion and epilimnion in stratified lakes as phosphorus transport processes. His mathematical formulation for the phosphorus budget in a lake did not include exchange at the sediment water interface, although he recognized its existence. In a later paper (Snodgrass and O'Melia, 1975), a more detailed model for phosphorus was presented in which orthophosphate and particulate phosphorus were considered separately. Separate mathematical formulations were presented for the epilimnion and hypolimnion for both forms of phosphorus. Internal loading of phosphorus from the sediments was still not included, although mineralization of organic phosphorus in the hypolimnion by heterotrophic bacteria was included. Sedimentation of particulate phosphorus was accounted for and the uptake of orthophosphate in biomass production was represented as proportional to the amount of orthophosphate in the epilimnion and was transferred to the particulate phosphorus material balance after conversion. It was suggested that flocculation of particulate phosphorus in the hypolimnion may be important in accelerating the sedimentation process.

Imboden (1974) described a similar two-box two-compartment model for phosphorus which incorporated phosphorus exchange at the sediment-water interface. Through physical reasoning and analysis of the time to achieve steady-state using the model, Imboden concluded that shallow lakes with short detention periods will respond relatively quickly to changes in phosphorous loading. This would be true of the western basin of Lake Erie in which the mean depth is 7.30 m and the detention period is 0.13 year. Conversely, deep lakes will respond very slowly to changes in phosphorus loading. However, phosphorus loading from sediments could alter the expected response of any lake if not properly accounted for, particularly when bottom waters become anaerobic for a significant period.

Lung et al. (1976) applied a two-box, two-compartment phosphorus model to a lake in southwest Michigan with a surface area of $1.05 \times 10^4 m^2$, a mean depth of 7.9 m, and a hydraulic detention period on the order of 1-2 months. The model is similar to the one presented by Snodgrass and O'Melia with the exception that the sediment is also modeled as a layered system incorporating both sedimentation of particulate phosphorus and diffusion of dissolved phosphorus in the sediments. They found their calibrated model to be a fairly good predictor of the phosphorus budget when the lake outflow increased about 30 percent in a subsequent year. This episode was used as a verification of the model. An unusual feature of this model was the simulation of particulate and dissolved phosphorus in the first 38 cm below the sediment-water interface. This was attempted because of the availability of sediment phosphorus data.

Chapra (1977) formulated a model for total phosphorus in the Great Lakes. Each lake was considered as a completely stirred system with the exception of Lake Erie which was divided into three sub-basins, each completely mixed. The model time step was one year and in-lake losses of phosphorus were modeled by a sedimentation process, $S = vA_sp$, where v represented the "apparent" settling velocity of total phosphorus, A_s was the surface area of the sediments, and p denoted the concentration of total phosphorus in the lake water. A value of 16 m/yr was selected for the apparent settling velocity. The total phosphorus level in each basin was simulated historically for the period 1800 to 1970 using estimated loadings based on available information for point and diffuse sources. Future conditions were simulated for hypothetical reductions in phosphorus loading. Chapra cited the neglect of phosphorus regeneration from the sediments and the underestimation of the magnitude of diffuse sources as the most serious qualifiers to his model output.

Lorenzen et al. (1976) formulated a material balance expression for total phosphorus in Lake Washington which incorporated sedimentation, regeneration of phosphorus from the sediments, and retention of phosphorus in the sediments. Rate constants were determined by calibrating the model for the period 1940 to 1950 and were not changed in the subsequent use of the model for the period 1950 to 1970 during which large changes in loading occurred. The model results were in good agreement with observed concentrations of total phosphorus during this period.

In an ongoing wastewater management study (U. S. Army Corps of Engineers, 1975), the Lorenzen et al. (1976) model was adapted for application to Lake Erie. This adaptation is described below.

The essential features of the model are depicted in Figure 3. In this figure, M represents the annual phosphorus loading to the basin in grams per year (g/yr). C is the phosphorus concentration in the lake water in grams per cubic meter (g/m^3). C_s is the phosphorus concentration in the interstitial water of the sediments in grams per cubic meter. K_1 is the specific rate of phosphorus transfer to the sediments in meters per year (m/yr). The product K_1C represents the annual phosphorus transfer to the sediments in grams per square meter. When this product is multiplied by the sediment surface area in square meters, the total annual phosphorus transfer to the sediments is obtained in grams per year. K_2 is the specific rate of phosphorus release from the sediments in meters per year. The annual phosphorus release per square meter of sediment area is given by the product K_2 and the difference in phosphorus concentration between the interstitial waters and the lake waters. K_3 is the fraction of the total phosphorus input to the sediment which is retained in the sediments and no longer available for exchange. Q denotes the outflow from the basin in cubic meters per year (m^3/yr). Inflow is assumed equal to outflow. The product CQ is the annual export of phosphorus from the basin via the outflow in grams per year.

The material balance equation for phosphorus in the lake water can be written as

$$\frac{dC}{dt} = \frac{M}{V} + \frac{K_2(C_s - C)A}{V} - \frac{K_1CA}{V} - \frac{CQ}{V} \qquad (1)$$

where V is the lake basin volume assumed constant in cubic meters and A is the sediment area in square meters. The quantity dC/dt represents the annual change in lake phosphorus concentration required to balance the phosphorus budget. If the basin is at equilibrium and the phosphorus loading is maintained constant, the lake phosphorus concentration remains unchanged and dC/dt equals zero.

The material balance equation for phosphorus in the sediment interstitial water is given by

$$\frac{dC_s}{dt} = = \frac{K_2(C_s - C)A}{V_s} + \frac{K_1CA}{V_s} - \frac{K_1K_3AC}{V_s} \qquad (2)$$

where V_s is the interstitial water volume. Again, for equilibrium conditions, the quantity dC_s/dt equals zero.

The above set of coupled equations can be solved for C and C_s as a function of time measured in years. The accuracy of the solution in quantitative terms depends on the adequacy of the model formulation, correctness of the values chosen for the three rate constants (K_1, K_2, and K_3), adequate information on the phosphorus loading rate (M), and data on the lake characteristics (V, V_s, A, and Q). Initial values of C and C_s must also be known. If equilibrium conditions are assumed such that dC/dt = 0, the following two equations are obtained which relate the three rate constants.

$$K_1K_3 = \frac{M}{AC_e} - \frac{Q}{A} \tag{3}$$

$$K_2 = K_1(1 - K_3)\frac{C_e}{C_{s_e} - C_e} \tag{4}$$

In this case, C_e and Cs_e are the steady-state or equilibrium values for lake and interstitial water concentrations respectively. With knowledge of any one of the rate constants, the other two are specified by the above equations (3) and (4).

The three basin version of this modeling approach for Lake Erie is depicted in Figure 4. In the notation, the first subscript refers to the basin (i.e., C_1 refers to the lake phosphorus concentration in the western basin) and the second subscript denotes the specific rate constant (i.e., K_{12} is the specific rate of phosphorus release, K_2, in the western basin, i.e., basin 1). The physical data for the three basins are given in Table IV along with the estimated phosphorus concentrations for the year 1975. Interstitial water phosphorus concentrations were measured in the central basin and an average value for the central basin was given as $Cs_2 = 0.286\ g/m^3$. Assuming that interstitial water phophorus concentrations varied in proportion to the total sediment phosphorus content as reported by the Federal Water Pollution Control Administration (1968), the values for the western and eastern basins were obtained and are given in Table IV.

The rate constants were obtained in the following manner. Based on the findings of Project Hypo (1972), a value of K_{22} for the central basin was determined for both oxic and anoxic conditions. The oxic phosphorus release constant was estimated to be 1.0 m/yr. This value was used for the western basin, the eastern basin, and in the case of the central basin, for 9 months of the year. The anoxic phosphorus release constant was estimated to be 11.0 m/yr. This latter value was assumed to apply in the central basin for 3 months of the year. The annual average values for the phosphorus release constants then become: $K_{12} = 1.0$ m/yr, $K_{22} = 3.5$ m/yr, and $K_{32} = 1.0$ m/yr.

The total external load of phosphorus to each basin is given by M_i, where i = 1, 2, or 3 for the western, central or eastern basin, respectively. The external load for each basin is given by the sum of the basin's tributary area load estimates and the outflow from the upstream basin. Estimates for the phosphorus loadings to the three basins are thus given by the equation $M_i = m_i + QC_{i-1}$, where m_i is the basin tributary area contribution and QC_{i-1} is the

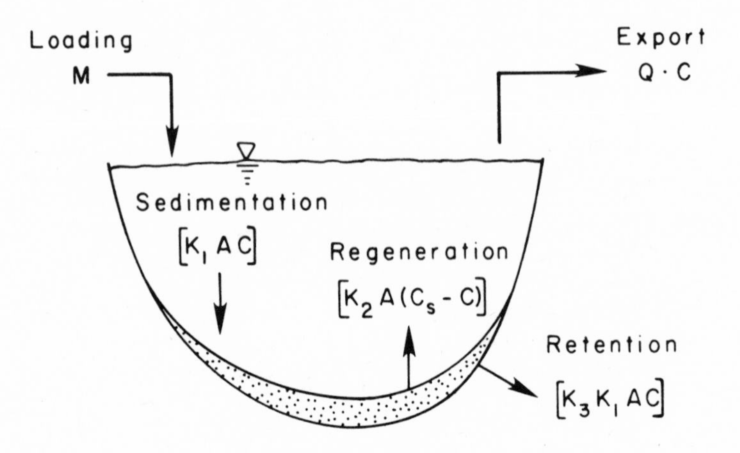

Figure 3. Schematic Representation of Transport Processes used in Phosphorus Budget Model (U. S. Army Corps of Engineers, 1975).

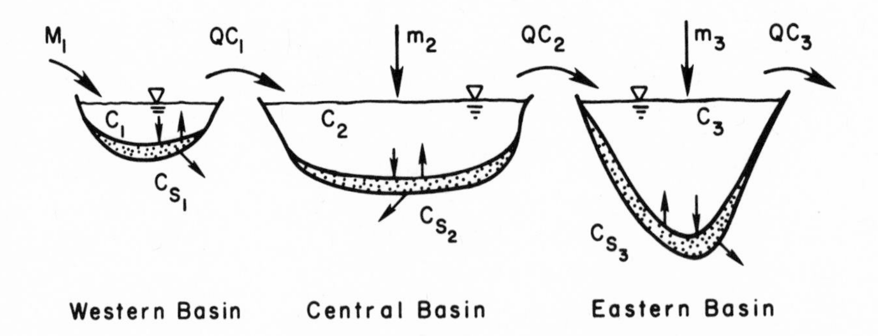

Figure 4. Three-Basin Phosphorus Budget Model for Lake Erie (U. S. Army Corps of Engineers, 1975).

Table IV

WESTERN BASIN	CENTRAL BASIN	EASTERN BASIN
$V_1 = 2.27 \times 10^{10} m^3$	$V_2 = 2.98 \times 10^{11} m^3$	$V_3 = 1.52 \times 10^{11} m^3$
$V_{s_1} = 3.11 \times 10^8 m^3$	$V_{s_2} = 1.63 \times 10^9 m^3$	$V_{s_3} = 6.24 \times 10^8 m^3$
$A_1 = 3.11 \times 10^9 m^2$	$A_2 = 1.63 \times 10^{10} m^2$	$A_3 = 6.24 \times 10^9 m^2$
$Q = 18.1 \times 10^{10} m^3/yr$	$Q = 18.1 \times 10^{10} m^3/yr$	$Q = 18.1 \times 10^{10} m^3 yr$
$*C_1 = .037 g/m^3$	$C_2 = .018 g/m^3$	$C_3 = .022 g/m^3$
$C_{s_1} = .336 g/m^3$	$C_{s_2} = .286 g/m^3$	$C_{s_3} = .221 g/m^3$

*Phosphorus concentrations are estimated 1975 values (U. S. Army Corps of Engineers, 1975).

flow times the phosphorus concentration in the upstream basin. The current estimates for the loads in each of the basins are (from Table II) given in Table V.

Table V

Estimates of Phosphorus Loadings to the Three Basins of Lake Erie, 1975

m_1	=	1.3281×10^{10} g/yr	(m_1	=	M_1)
m_2	=	0.4520×10^{10} g/yr	(M_2^*	=	1.1217×10^{10} g/yr)
m_3	=	0.1806×10^{10} g/yr	(M_3^*	=	0.5064×10^{10} g/yr)

M_2 and M_3 include phosphorus exported from upstream basin.

If it is assumed that equilibrium conditions prevailed during the year 1975, the other two rate constants, K_1 and K_3 for each basin, can be estimated using equations (3) and (4). They are as follows:

K_{11} = 65.30 m/yr, K_{13} = 0.876

K_{21} = 79.24 m/yr, K_{23} = 0.342

K_{31} = 16.93 m/yr, K_{33} = 0.466

With the above information, the phosphorus budget model was utilized to examine the consequences of reduced phosphorus loadings on the lake phosphorus concentrations and interstitial water phosphorus concentrations. The model depicts the transient response of each basin as well as the new equilibrium phosphorus concentrations after a change in external loading. The transient concentrations will have some sensitivity to the rate constants, although they are primarily dependent upon basin volumes and the water discharge.

PHOSPHORUS RETENTION

The above described model would appear to incorporate the principal transport mechanism for total phosphorus for a lake basin in which exchange of phosphorus with the sediments is important. However, ignoring the internal phosphorus exchange processes represented by the rate constants K_1, K_2, and K_3 leads to a very simple mass balance equation for equilibrium conditions (i.e., when $C = C_e$, equilibrium concentration). This can be done by defining a phosphorus retention coefficient, R, given by

$$R = 1 - \frac{QC_e}{M} \tag{5}$$

The retention coefficient represents the proportion of the external loading of phosphorus that is retained in the lake, generally over a one-year period. Using the notation from the model presented above, the retention coefficient can also be given by

$$R = \frac{K_3 K_1 A C_e}{M} \tag{6}$$

in which the numerator represents that part of the sedimented total phosphorus which is retained in the sediment matrix.

As Chapra and Tarapchak (1976) have noted, equation (5) is written for equilibrium conditions and for nonequilibrium conditions it should be written as:

$$R = 1 - \frac{QC}{M} - \frac{V}{M}\frac{dC}{dt} \tag{7}$$

in which the last term represents the possible error in estimating R if equilibrium conditions have been assumed when they do not exist.

Kirchner and Dillon (1975) correlated the phosphorus retention coefficient, R, with areal water loading, q_s, for several lakes and obtained the following equation:

$$R = 0.426 \exp(-0.271 q_s) + 0.574 \exp(-0.00949 q_s) \tag{8}$$

Larsen and Mercier (1976), in a similar type of analysis, found the phosphorus retention coefficient to be correlated with either the areal water loading, q_s, or the flushing rate, $\rho = Q/V$. However, the lakes selected for analysis were primarily oligotrophic.

Equation (3) can be rewritten as follows by letting $K_1K_3 = v$, the apparent settling velocity (Chapra, 1975), and $M/A = L$, the areal phosphorus loading.

$$C_e = \frac{L}{v + q_s} \tag{9}$$

Substitution of equation (9) into equation (5) gives

$$R = \frac{v}{v + q_s} \tag{10}$$

Equation (10) has the desired characteristics of $R \rightarrow 1$ as $q_s \rightarrow 0$ and $R \rightarrow 0$ as $q_s \rightarrow \infty$. Both of these limiting conditions are consistent with physical reasoning. Chapra (1977) assigned a value of $v = 16$ m/yr for the Great Lakes. The computed values for the three basins of Lake Erie (using $v = K_1K_3$) show wide variability as follows: $v_1 = 57.2$ m/yr, $v_2 = 29.1$ m/yr, and $v_3 = 7.89$ m/yr. An average value for the whole lake can be computed using equation (9) which gives $v = 31$ m/yr.

Various estimates of the retention coefficients for Lake Erie are summarized in Table VI.

Table VI

Estimated Phosphorus Retention Coefficients for Lake Erie

Source	Western Basin R_1	Central Basin R_2	Eastern Basin R_3	Whole Lake R
Corps of Engineers (1975)	0.50	0.71	0.21	0.81
Burns *et al*. (1976)	0.74	0.88	0.40	0.91
Kirchner and Dillon (1975) [Equation (8)]	0.33	0.54	0.44	0.60
Chapra (1977)	0.22	0.59	0.36	0.69

The variability in the estimated values for the phosphorus retention coefficient for Lake Erie raises a question regarding the assumption of equilibrium conditions. The possible error in computing R for a given basin can be estimated by assuming values for dC/dt in equation (7). For example, if in 1975 the assumption of equilibrium conditions was incorrect and $dC/dt = +.005$ g/m^3/yr, then the corrected values for the retention coefficients (Corps of Engineers, 1975) would be $R_1 = 0.49$, $R_2 = 0.58$, $R_3 = 0.06$, and $R = 0.69$. It can be readily seen that significant errors are possible. Because of the relatively low value of V/M for the western basin, the possible error in computing R is smallest in that basin. However, the application of the phosphorus retention concept as presented using equations (5) or (7) to the sub-basins of Lake Erie is dubious because of the intermittent exchange of water mass between the basins during seiching and when winds originate from the east. This occaisional reverse (or backward) flow of water mass from downstream basin to upstream basin may be a significant factor in the levels of phosphorus found in the basins under certain conditions. Measurements taken at such time would not be representative of a single basin equilibrium phosphorus condition as represented by equation (5).

TROPHIC STATUS

Equation (5) can be rearranged after introduction of the mean depth, $z = V/A$, to give

$$\frac{L(1 - R)}{\rho z} = C_e \tag{11}$$

which is the basis for the trophic status graph introduced by Vollenweider (1971) and modified by Dillon (1975). Figure 5 shows Lake Erie conditions plotted on the trophic status graph. This type of analysis indicates that the western and eastern basins are eutrophic, the central basin is marginally eutrophic, and the whole lake is eutrophic.

An alternate trophic status graph can be obtained by solving equation (9) for the unit area phosphorus loading, L;

$$L = C_e(v + q_s) \tag{12}$$

Using the critical phosphorus concentrations, $C_e = .01\ g/m^3$ (dividing oligotrophic and mesotrophic lakes) and $C_e = .02\ g/m^3$ (dividing mesotrophic and eutrophic lakes), and the settling velocity, v = 16 m/yr, given by Chapra (1977), the following equations are obtained for permissable and dangerous area phosphorus loadings:

$$L_{permissable}\ (g/m^2/yr) = 0.01(16 + q_s) \tag{13}$$

$$L_{dangerous}\ (g/m^2/yr) = 0.02(16 + q_s) \tag{14}$$

These equations along with the data from the Corps of Engineers (1975) are shown in Figure 6. The trophic status of Lake Erie is seen to be about the same for both methods of classification. This is reasonable since R and q_s have been demonstrated to correlate (Kirchner and Dillon, 1975; Larsen and Mercier, 1976).

The trophic recovery of Lake Erie is assumed to be primarily dependent upon reducing the phosphorus loading to the western basin. The use of a phosphorus budget model and a trophic status graph provides a method for estimating the in-lake effects that might result from a phosphorus control program. This approach does not provide information on the exact water quality at any specific point in time nor at a specific point in space within a basin. There are numerous drawbacks to this approach; however, public opinion and governmental legislation call for action.

The 1972 U. S.-Canada Great Lakes Water Quality Agreement

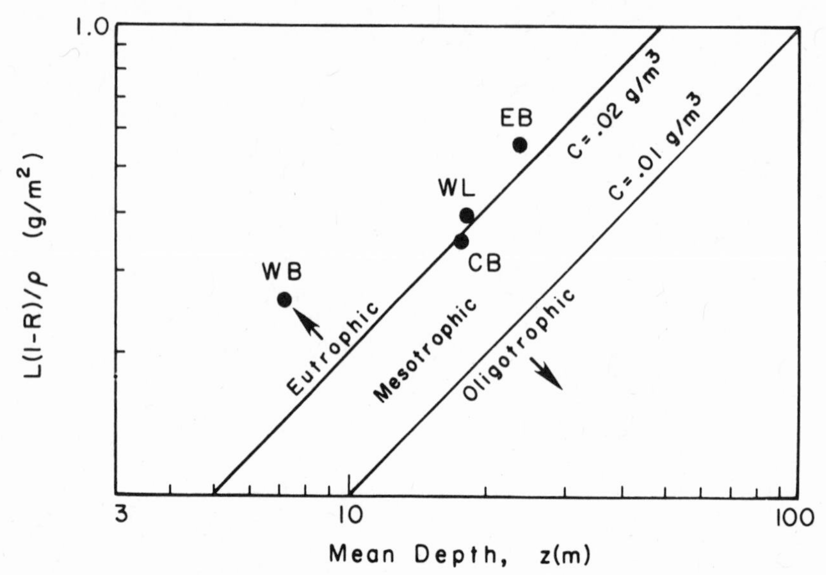

Figure 5. Trophic Status of Lake Erie Sub-Basins. o, Estimated Trophic Status, 1975 (U.S. Army Corp of Engineers); WB, Western Basin; CB, Central Basin; EB, Eastern Basin; WL, Whole Lake.

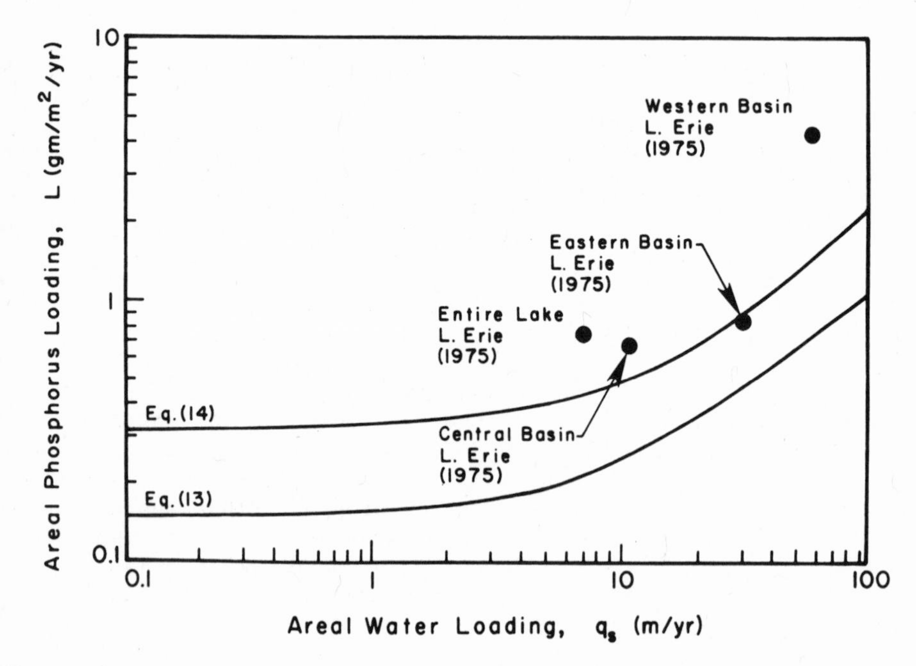

Figure 6. Areal Phosphorus Loading versus Areal Water Loading (after Chapra and Tarapchak, 1976).

gives the following objective for phosphorus:

> "Concentration should be limited to the extent necessary to prevent nuisance growths of algae, weeds, and slimes that are or may become injurious to any beneficial water use."

The International Joint Commission's Great Lakes Water Quality Board report of 1974 recommends that this objective be retained and further states:

> "The existing specific objective is in narrative form because the variable response of aquatic organisms dependent in part on phosphorus to produce nuisance conditions makes selection of a defensible single number very difficult."

Based on the trophic categorization used in this study, this objective would lead to average total phosphorus concentrations of 0.020 mg/l or less in each of the Lake's sub-basins. PL 92-500 (1972) calls for the "rehabilitation and environmental repair" of Lake Erie. Reduction of in-lake total phosphorus concentrations to 0.02 mg/l or below will rehabilitate the Lake to a more desirable condition, but the degree of rehabilitation is not strictly quantifiable. However, when historical phosphorus loading estimates are compared with current ones and when the present relative conditions of the three sub-basins are compared, the reasonable assumption can be made that the central and eastern basins historically have had lower phosphorus concentrations than the western basin. It was for that reason that the U. S. Army Corps of Engineers (1975) selected the following objective total phosphorus concentrations: western basin, $C_{e_1} = 0.02\ g/m^3$; central basin, $C_{e_2} = 0.015\ g/m^3$; and eastern basin, $C_{e_3} = 0.015\ g/m^3$.

The total phosphorus concentration objectives used for the Lake Erie Study were defined because it was necessary to have some quantitative goal to use in determining required phosphorus load reductions.

PHOSPHORUS MANAGEMENT

The proposed Lake Erie Study objective total phosphorus concentrations can be used in equations (11) and (12) to develop estimates of the reduction in phosphorus loadings required to achieve these new objective equilibrium concentrations. The results are shown in Table VII.

It is the objective of the U. S.-Canada Great Lakes Water Quality Agreement of 1972 to reduce phosphorus concentrations from

Table VII

New Phosphorus Loadings to Lake Erie to Achieve Objective Phosphorus Concentrations

	Objective Phosphorus Concentration C_e	Areal Phosphorus Loading Eq. (11)	Areal Phosphorus Loading Eq. (12)	Tributary Loadings M (from Eq. (11))	Tributary Loadings M (from Eq. (12))
Western Basin	.02 g/m^3	2.33 $g/m^2/yr$	1.48 $g/m^2/yr$	4912	2269
Central Basin	.015 g/m^3	0.577 $g/m^2/yr$	0.407 $g/m^2/yr$	5785	3000
Eastern Basin	.015 g/m^3	0.55 $g/m^2/yr$	0.675 $g/m^2/yr$	717	1497

all municipal plants with capacities of one million gallons per day or greater to less than 1 g/m^3. The present Lake Erie situation (Corps of Engineers, 1975) and the new situations called for in Table VII are depicted schematically in Figures 7, 8, and 9.

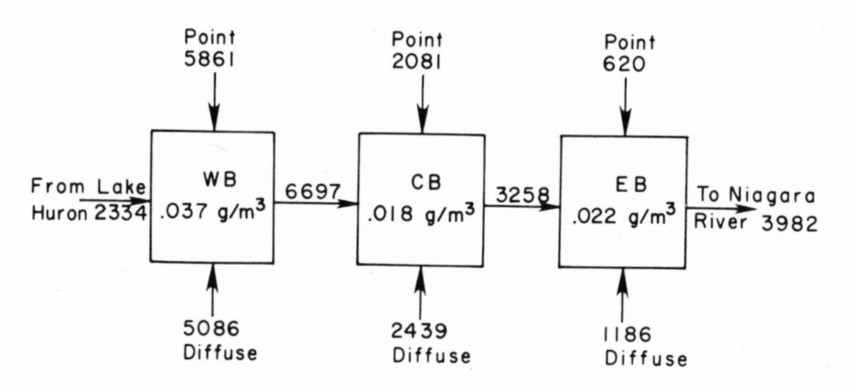

Figure 7. Estimated Phosphorus Loads and Concentrations in Lake Erie (1975) (Loads in metric tons/year).

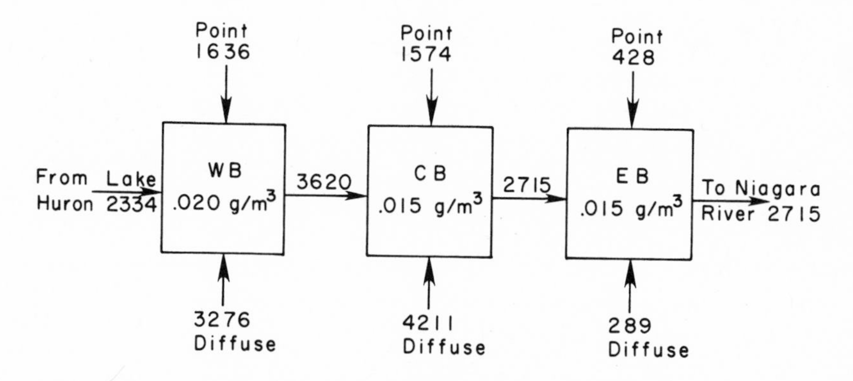

Figure 8. PLAN I: Projected Phosphorus Loads from Eq. (11) and Objective Concentrations in Lake Erie (Loads in metric tons/year).

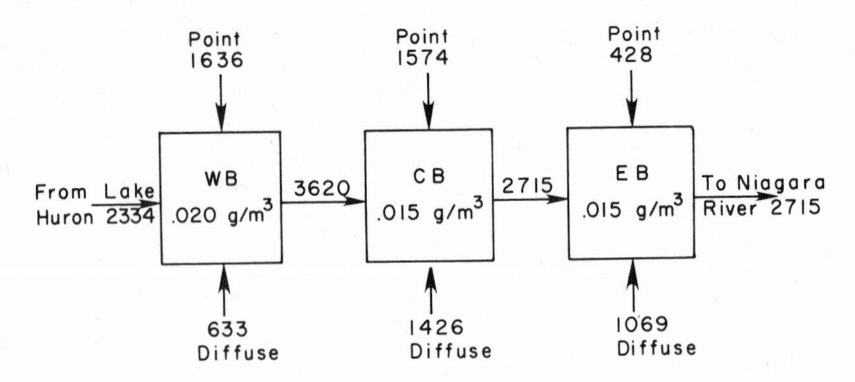

Figure 9. PLAN II: Projected Phosphorus Loads from Eq. (12) and Objective Concentrations in Lake Erie (Loads in metric tons/year).

The projected diffuse phosphorus loads shown in Figures 8 and 9 differ markedly. In the case of the plan presented in Figure 8, the greatest uncertainty is in the magnitude of the phosphorus retention coefficients used. In the case of the plan presented in Figure 9, the greatest uncertainty is in the magnitude of the apparent settling velocity used. Both plans call for the same point source loadings (i.e., effluent phosphorus concentrations of 1 g/m^3 for waste discharges greater than 1 mgd). The two plans call for the following reductions in diffuse source loading.

Table VIII

Percent Reduction in Diffuse Source Loading

	Western Basin	Central Basin	Eastern Basin
Plan I	36%	no reduction	76%
Plan II	88%	42%	10%

These two possible management alternative plans which are based on existing point source loading reduction objectives indicate that although management of point sources alone will have a beneficial impact on the water quality of Lake Erie, it will not sufficiently improve the trophic conditions in all basins of the lake. Reduction of diffuse source loads will also be required.

SUMMARY

There exist an infinite number of alternatives when considering reduction of both point source and diffuse source loading to the three basins in Lake Erie. As improved information on pollutant loads and in-lake conditions becomes available, it will be possible to refine the estimates of in-lake phosphorus concentrations that would result from specific load reduction plans. Before implementing such plans, their effect on the activities in the watersheds involved will need to be evaluated, since load reductions will only be achieved after appropriate source management is instituted. Impacts of possible management techniques, such as modified tillage practices, reductions of the phosphorus content in fertilizers, and sediment catchment basins must be weighed against the expected improvement of the lake. Phosphorus removal from municipal and industrial point sources is already underway. The challenge ahead will be to significantly reduce the diffuse source loading.

Although the overall strategy to improve the quality of Lake Erie and thereby achieve a more desirable trophic status seems clear, the details of phosphorus movement and changes in form is not as clear. The coupling of time-dependent hydrodynamic models with time-dependent phytoplankton models is one important avenue to developing a better understanding of these details. With such understanding of the biochemical reactions and hydrodynamic transport, it will be possible to more accurately quantify the phosphorus budget and predict the response of the Lake Erie ecosystem to changes in phosphorus loading. This will all come slowly because of limitations in computational capability and the cost of obtaining field data for calibration and verification of multi-compartment simulation models.

ACKNOWLEDGEMENTS

In addition to the authors and sources cited, many other individuals have aided me in gaining a better understanding of phosphorus transport in Lake Erie. For this, I am grateful to Angelo Coniglio, Stephen Yaksich, and David Melfi of the Lake Erie Wastewater Management Study staff, U. S. Army Corps of Engineers, Buffalo District. I also wish to thank Akio Wake, State University of New York at Buffalo, and Randy Crissman, University of Delaware, for their valuable assistance. The patience and expert typing of Ms. Janis DeLussey is gratefully appreciated.

REFERENCES

Burns, N. M., J. D. H. Williams, J.-M. Jaquet, A. L. W. Kemp, and D. C. L. Lam, 1976a, "A Phosphorus Budget for Lake Erie," Journal of the Fisheries Research Board of Canada, Vol. 33, No. 3, pp. 564-573.

Burns, N. M., 1976b, "Nutrient Budgets for Lake Erie, 1970," Journal of the Fisheries Research Board of Canada, Vol. 33, No. 3, pp. 520-536.

Burns, N. M., 1976c, "Temperature, Oxygen, and Nutrient Distribution Patterns in Lake Erie, 1970," Journal of the Fisheries Research Board of Canada, Vol. 33, No. 3, pp. 485-511.

Chapra, S. C., 1977, "Total Phosphorus Model for the Great Lakes," Proceedings of the American Society of Civil Engineers, Vol. 103, No. EE2, pp. 147-162.

Chapra, S. C. and S. J. Tarapchak, 1976, "A Chlorophyll a Model and Its Relationship to Phosphorus Loading Plots for Lakes," Water Resrouces Research, Vol. 12, No. 6, pp. 1260-1264.

Dillon, P., 1975, "The Phosphorus Budget of Cameron Lake, Ontario: The Importance of Flushing Rate to the Degree of Eutrophy of Lakes," Limnology and Oceanography, Vol. 20, No. 1, January, pp. 28-39.

Dillon, P., 1975, "The Application of the Phosphorus Loading Concept to Eutrophication Research," Scientific Series No. 46, Canada Centre for Inland Waters, Ontario.

DiToro, D. M., D. J. O'Connor, J. L. Mancini, and R. V. Thomann, 1973, "A Preliminary Phytoplankton-Zooplankton-Nutrient Model of Western Lake Erie," In Systems Analysis and Simulation in Ecology, Vol. 3, 592 p.

Freedman, P. L. and R. P. Canale, 1977, "Nutrient Release from Anaerobic Sediments," Proceedings American Society of Civil Engineers, Vol. 103, No. EE2, pp. 233-244.

"Great Lakes Water Quality Agreement, with Annexes and Texts and Terms of Reference, Between the United States of America and Canada," April 15, 1972.

Hornberger, G. M., M. G. Kelly, and T. C. Lederman, 1975, "Evaluating a Mathematical Model for Predicting Lake Eutrophication," Bulletin 82, Virginia Water Resource Research Center, Virginia Polytechnic Institute and State University.

Hutchinson, G. Evelyn, 1957, A Treatise on Limnology, Vol. 1, John Wiley and Sons, Inc., New York.

Imboden, D. M., 1974, "Phosphorus Model for Lake Eutrophication," Limnology and Oceanography, Vol. 19, No. 2, pp. 297-304.

Kirchner, W. B. and P. J. Dillon, 1975, "An Empirical Method of Estimating the Retention of Phosphorus in Lakes," Water Resources Research, Vol. 11, No. 1, pp. 182-183.

"Lake Erie Environmental Summary: 1963-64," U. S. Department of Interior, Federal Water Pollution Control Administration, 1968.

Larsen, D. P. and H. T. Mercier, 1976, "Phosphorus Retention Capacity of Lakes," Journal of the Fisheries Research Board of Canada, Vol. 33, No. 8, pp. 1742-1750.

Lorenzen, M. W., D. J. Smith, and L. V. Kimmel, 1976, "A Long-Term Phosphorus Model for Lakes: Application to Lake Washington," in Modeling Biochemical Processes in Aquatic Ecosystems, edited by R. P. Canale, pp. 75-92, Ann Arbor Science, Ann Arbor, Michigan.

Lung, W. S., R. P. Canale, and Paul L. Freedman, 1976, "Phosphorus Models for Eutrophic Lakes," Water Research, Vol. 10, pp. 1101-1114.

O'Melia, C. R., 1972, "An Approach to the Modeling of Lakes," Schweizerische Zeitschrift fur Hydrologie, Vol. 34, pp. 1-33.

"Project Hypo--An Intensive Study of the Lake Erie Central Basin Hypolimnion and Related Surface Water Phenomena," Canada Centre for Inland Waters and U. S. Environmental Protection Agency, Region V, 1972.

Richards, F. A., J. D. Cline, W. W. Broenkow, and L. P. Atkinson, 1965, "Some Consequences of the Decomposition of Organic Matter in Lake Nitinat, An Anoxic Fjord," Limnology and Oceanography, Vol. 10, pp. 185-201.

Sager, Paul, 1976, "Ecological Aspects of Eutrophication," Water International, Vol. 1, No. 2, pp. 11-17.

Snodgrass, W. J. and C. R. O'Melia, 1975, "Predictive Model for Phosphorus in Lakes," Environmental Science and Technology, Vol. 9, No. 10, pp. 937-944.

Thomann, R. V., R. P. Winfield, and D. M. DiToro, 1974, "Modeling of Phytoplankton in Lake Ontario," Proc. 17th Conference on Great Lakes Research, International Association for Great Lakes Research, pp. 135-149.

United States Army Corps of Engineers, 1975, "Lake Erie Wastewater Management Study, Preliminary Feasibility Report," 3 volumes, Corps of Engineers, Buffalo District, Buffalo, New York.

Vollenweider, R. A., 1971, "Scientific Fundamentals of the Eutrophication of Lakes and Flowing Waters, with Particular Reference to Nitrogen and Phosphorus as Factors in Eutrophication," Organization for Economic Cooperation and Development, Paris, France.

Water Pollution Control Act, Amendments of 1972, Public Law 92-500, 92nd Congress, S. 2770, U. S. Congressional Directory.

Williams, J. D. H., J.-M. Jaquet, and R. L. Thomas, 1976a, "Forms of Phosphorus in the Surficial Sediments of Lake Erie," Journal of the Fisheries Research Board of Canada, Vol. 33, pp. 413-429.

OXYGEN TRANSFER AT THE AIR-WATER INTERFACE

E. R. Holley

University of Illinois at Urbana-Champaign

ABSTRACT

This review paper is primarily concerned with the mechanics of oxygen absorption across the free surface of natural bodies of water. A summary is given of basic concepts related to the solution of oxygen in water, to the rate at which oxygen is absorbed by water, and to the manner in which turbulence in the water influences the absorption or reaeration process. A review is given of some of the analytical models which have been proposed, and some of the empirical relationships which have been derived, for representing the reaeration process. References are cited for several critical reviews of various expressions which have been proposed in the literature for relating the oxygen transfer coefficient, K_L, to hydraulic conditions and to fluid properties for flow in streams. These critical reviews indicate that there are significant differences between the prediction equations. Some of these differences apparently result from the fact that all of the variables which have a significant effect on K_L have sometimes not been considered in developing the prediction equations. Some of the potentially significant aspects which have not always been given adequate consideration are the boundary layer nature of the surface film and the influence of suspended sediment and wind on reaeration. The available laboratory and field data are presented for the effects of wind speed on K_L for both streams and larger bodies of water with small velocities. There is good agreement among the various sets of laboratory data, but there are significant differences between the laboratory data and the space field data on wind effects on K_L.

1. INTRODUCTION

1.1 General Problem

Oxygen transfer at the air-water interface is part of the general problem of gas absorption by liquids. This general problem has been studied by scientists and engineers in several disciplines (e.g. chemical engineers, fermentation technologists, civil engineers, etc.); much can be learned about this process from work done on similar problems by persons in several disciplines.

The problem of oxygen transfer at the air-water interface is of particular interest since the dissolved oxygen content of water has been, and no doubt will continue to be, a major water quality parameter. Oxygen absorption by water is often called aeration or reaeration.

There are several ways in which the study of reaeration can be subdivided. One possible division is between reaeration in natural water bodies (e.g. in streams, estuaries, lakes, oceans) and reaeration which is not in a natural environment (e.g. in a wastewater treatment facility). Another type of division would be between natural and artificial reaeration. Natural reaeration refers to the oxygen transfer across the natural free surface while artificial reaeration is normally concerned with augmenting the natural transfer by creating additional mixing, additional interfacial area (e.g. by bubbles), and/or additional partial pressure of the oxygen in the gaseous phase.

The general topic of oxygen transfer at the air-water interface is quite broad and could be approached from several viewpoints. No attempt will be made here to cover, or even to cite references on, all aspects of the problem. Attention will be focused on natural reaeration and emphasis will be given primarily to reaeration in relation to stream flow characteristics. In Section 5, the influence of wind on reaeration will be considered.

The primary objectives of this paper are to review briefly some basic concepts in reaeration and some of the past work on models and prediction equations for streams, to present some considerations relative to boundary layer concepts and the possible influence of suspended particles on reaeration, and to summarize some of the data on wind effects. More comprehensive reviews of the literature on prediction equations for reaeration rate coefficients in streams are cited in Section 2.4. Investigations have been conducted on many aspects of reaeration which will not be considered here, for example, bubble and spray reaeration, the influence of surface active agents and of surface contaminants, self-aerated or air-entraining flows, reaeration induced by flow through various types of hydraulic structures, oxygen balances and distributions in rivers, lakes, reservoirs, and estuaries, methods for determining reaeration rate coefficients, various types of artificial reaeration devices, the influence of chemical and biological reactions, etc. The Journal of the Water Pollution Control Federation normally includes a section on Oxygen Sag and Stream Purification in their annual literature review.

Reaeration takes place because there is a saturation concentration at which the dissolved oxygen in water is at equilibrium with the atmosphere to which the water is exposed and because the dissolved oxygen content of water naturally tends to this equilibrium. The saturation concentration (c_s) is given by Henry's Law which can be stated (Sawyer and McCarty, 1967) as

$$c_s = \alpha' p \tag{1}$$

where p is the partial pressure of oxygen and α' is the Henry's Law constant. The value of α in Eq. 1 depends on temperature and on the characteristics of the water. Values of c_s for distilled water were given by the ASCE Committee on Sanitary Engineering Research (1960). Those values of c_s should not be considered as universal since c_s can vary with the characteristics of the water. For example, Holley et al. (1970) found that c_s values for tap water could be on the order of 1 mg/l lower than those reported by the Committee. Other investigators have sometimes found similar variations. Morris et al. (1961) concluded that variations in c_s from tabulated values could be on the order of 1%.

In natural water bodies, the water may be at a concentration less than the saturation value for many reasons. For example, the water may be at, or may have been released from, the lower levels of a reservoir or biological decomposition of organic material may have exerted an oxygen demand which utilized dissolved oxygen. If the dissolved oxygen concentration (C) is less than c_s, oxygen is then absorbed causing C to tend naturally to the saturation or equilibrium value. One of the primary areas of interest in the study of oxygen budgets is the rate at which this natural reaeration takes place.

During reaeration, oxygen molecules are both entering and leaving the water continuously. If the concentration (C) of dissolved oxygen is less than saturation, the number of molecules entering per unit time is greater than the number leaving, so that there is a net increase in C. At saturation, the rates at which molecules enter and leave the water are equal so that the dissolved oxygen concentration is at equilibrium.

1.2 Reaeration Rate

In a mixing vessel, the rate of change of concentration (C) of dissolved oxygen in water with no oxygen demand has been found to be (Fair et al., 1968)

$$\frac{dC}{dt} = K_2 \ (c_s - C) \tag{2}$$

where K_2 is a reaeration rate coefficient having dimensions of $(\text{time})^{-1}$. The quantity $(c_s - C)$ is the saturation deficit, D. If c_s is constant, then Eq. 2 may be written as

$$\frac{dD}{dt} = -K_2 D \tag{3}$$

For a given deficit, D, the reaeration coefficient K_2 represents the net rate of oxygen absorption per unit volume of water.

For water with an initial deficit D_o at t=0, the solution to Eq. 3 is

$$\frac{D}{D_o} = e^{-K_2 t} = 10^{-k_2 t} \tag{4}$$

where $k_2 = K_2/\ln 10 = K_2/2.30$. In Eq. 4, the deficit approaches zero asymptotically (i.e., the concentration of dissolved oxygen approaches saturation asymptotically). The net rate of oxygen transfer decreases exponentially with time as the deficit approaches zero (Eqs. 3 and 4).

Equations 2 through 4 are given with time as the independent variable; in most cases, time may be interpreted as flow time for application to streams. This interpretation is equivalent to the use of a plug flow model which neglects longitudinal mixing. A plug flow model may not be valid if longitudinal dispersion is a significant transport mechanism in the problem being considered (Dobbins, 1964). Also, a plug flow model implies that the dissolved oxygen is uniformly distributed across each cross section in a stream. Churchill et al. (1962) observed rather consistent transverse variations of dissolved oxygen concentrations in their field tests. Rood and Holley (1974) showed that for some situations there are significant transverse variations of dissolved oxygen concentrations as a result of transverse mixing in streams. Eheart (1975) also considered the effects of transverse mixing on dissolved oxygen distributions.

Much research on reaeration has centered on being able to predict K_2 (or alternatively K_L in Section 2.1) for various flow conditions. In general, the coefficient K_2 is dependent on 1) the hydraulic characteristics of the stream flow, 2) wind and waves, 3) the Schmidt number (Sc), where $Sc = \nu/D_m^*$ = kinematic viscosity of the liquid divided by the molecular diffusivity of the dissolved gas, 4) any chemical reactions which may be taking place, and 5) any surface active agents which may be present. In addition, Tsao (1968) assumed that micro-organisms in water could absorb oxygen directly from the atmosphere. (See Section 4.) As will be reviewed in later sections, the rate controlling process in natural reaeration is usually the transport of dissolved oxygen away from the free surface in the liquid phase rather than the actual absorption of oxygen molecules across the gas-liquid interface. The first four items in the list above affect primarily the rate at which dissolved oxygen is transported downward from the free surface. The fifth item affects the penetration of oxygen molecules through the free surface. Oil films and other surface contaminants may also affect the reaeration process.

In problems involving the absorption of only one gas into water (as is the case in the reaeration problem), the dependence of K_2 on Sc may be replaced by a dependence on temperature (T) since Sc is a function primarily of T. The temperature dependence of K_2 on T is normally represented as

$$\frac{(K_2)_{T_2}}{(K_2)_{T_1}} = \theta^{(T_2 - T_1)} \tag{5}$$

Metzger (1968) has shown that the numerical value of θ depends on the mixing conditions in the water, with values being generally in the range 1.005 to 1.030. This variation of θ with mixing conditions is consistent with unpublished experimental work of the writer and with the observation in Section 3 that the boundary layer or film thickness (defined in Section 2) should be a function of both the Schmidt number (Sc) and the mixing conditions, as normally represented by the Reynolds number (Re) and relative roughness for shear flows. Sometimes, single values of θ are quoted in the literature without recognizing the possible variation of θ. Churchill et al. (1962) found θ to be the same for the two mixing conditions which they investigated.

Because of the influence of temperature on K_2, there is interest in the effects of thermal discharges on the oxygen economy of streams and other natural water bodies. See, for example, Edinger (1969), Fan and Hong (1972), Keshaven et al. (1973), Sornberger and Keshavan (1973), and Lin et al. (1975).

As mentioned previously, the dependence of K_2 on chemical reactions, surface active agents, artificial reaeration, etc. is not considered here.

1.3 Physical Process in Turbulent Water

The literature sometimes still refers to the oxygen transfer process as occurring through a supposedly stagnant (i.e. non-turbulent) film or layer at the free surface. However, there is little possibility of a stagnant film existing at the free surface of turbulent water. Rather, there is evidence to support the presence of turbulence at the free surface. Orlob (1959) and Engelund (1969) both used the free-surface turbulence in a wide channel as an example of homogeneous turbulence in which to study diffusion. Holley (1970) measured the one-dimensional spectrum of turbulence at distances from 0.006 in. (0.15 mm) to 1.0 in. (25 mm) below the free surface and found the same turbulence characteristics throughout this region. Approximate calculations indicated that some of these turbulence measurements were made in the region of the oxygen surface film, which is discussed in Section 2. Mattingly (1977) also made turbulence measurements near the free surface, but wind generated waves prohibited reliable measurements immediately below the surface. It is true that these various experiments dealt with the horizontal components of the turbulence while it is primarily the vertical component which is important in the problem of gas absorption from the atmosphere. Nevertheless, due to the essential three-dimensional nature of turbulence, there is no doubt that vertical turbulent motion as well as horizontal motion was present in the flow immediately below the free surface. There is a need for more detailed study of the nature of the turbulent flow field in a liquid very close to a gas-liquid interface.

The physical reaeration process may be viewed generally as follows:

> In turbulent water there is a random motion of the water. This random motion carries different parcels of water to the free surface at different times (Fig. 1A). If a parcel of water is below saturation concentration, then oxygen from the atmosphere is absorbed at the surface and moves into this parcel by the process of molecular diffusion during the time that the parcel is at the surface. This parcel, with its newly acquired dissolved oxygen, then moves back into the bulk of the liquid and is mixed with the surrounding water.

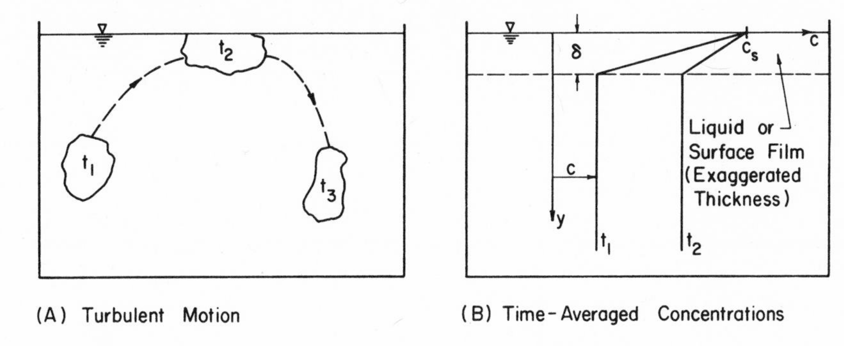

Fig. 1. Schematizations for the Reaeration Process.

Of course, this is an oversimplification since there is no parcel of water which remains completely intact even for a short exposure to the free surface.

In seeking to develop analytical models for reaeration, at least two alternatives are available: 1) The model can seek to describe the statistical nature of the turbulent mixing and the corresponding oxygen transfer while various water particles are exposed to the free surface. The models summarized in Section 2.2 are basically of this type. 2) The model can be based on the time-averaged value of the concentration distribution, which is approximately as shown in Fig. 1B. Both analytically (Holley et al., 1970) and experimentally (Churchill et al., 1962), it has been established for unstratified turbulent flows that the time-averaged oxygen concentration is essentially uniform over the depth (with the exception of a thin, film region immediately below the free surface). It seems logical to assume that the surface of the liquid is at saturation concentration based on the fact that (a) the ratio of the number of gaseous oxygen molecules bombarding the free surface to the net number going into solution is on the order of 10^8:1 (Miyamoto, 1931), (b) the time required to saturate a newly created surface is on the order of 10^{-7} seconds (Pasveer, 1953), and (c) models based on this assumption give results consistent with observed rates of reaeration. According to this assumption and the uniform distribution of dissolved oxygen over the depth, there must be a very thin region immediately below the free surface in which there is a steep gradient of dissolved oxygen concentration. Kanwisher (1963A) discussed some physical evidence for the existence of such a region. Holley (1973) has demonstrated the existence of such a region in the analogous problem of heat transfer by direct measurement of temperature distributions similar to the concentration distribution shown in Fig. 1B.

One purpose of seeking analytical reaeration models is to try to understand and represent the physical process with the ultimate objective of developing equations which can be used to predict reaeration rates for known flow conditions. As an alternative to analytical models (which generally look in varying detail at the mechanics of the reaeration process), it is possible to take an empirical or semi-empirical approach to seeking correlations between flow parameters and reaeration. Even the analytical models normally require empirical evaluation of some coefficients, parameters, or other unknown terms included in the models.

2. REVIEW OF SOME LITERATURE

The purpose of this section is to review some of the models and concepts which have been proposed in the literature for representing reaeration rate coefficients. Several references which have given comprehensive literature reviews are cited in Section 2.4.

2.1 Film Theory

Film theory represents one of the analytical approaches that has been used to study the absorption of gases into turbulent water and is based on the time-averaged concentration distribution. Film theory was first presented by Whitman (1923) and by Lewis and Whitman (1924) and states that the rate of absorption is controlled by the rate of diffusion through a gas film and a liquid film at the gas-liquid interface. The "films" are not physical films; rather they are thin layers or regions immediately adjacent to the interface. For gases (such as oxygen) having low solubility in water, the resistance to transfer through the gas film is negligible. Therefore, the dissolved oxygen concentration at the free surface is essentially in equilibrium with the oxygen partial pressure in the atmosphere (or is essentially equal to the saturation concentration, c_s) and the rate of absorption is controlled by diffusion through only the water film. Turbulence is assumed to keep the water vertically mixed below the film so that the concentration is uniform everywhere except in the film. Thus, through the thin film, a steep concentration gradient is assumed to exist, and the concentration distribution is as depicted in Fig. 1B.

Lewis and Whitman assumed that the transport through the film was solely by molecular action and that it therefore followed Fick's first law (Bird et al., 1960). Assuming the concentration distribution to be linear through the film of thickness δ, Fick's law for transport through the film was written as

$$q = AD_m^* \frac{c_s - C}{\delta} \quad (6)$$

where q is the rate of mass transport through the surface area A, C is the concentration below the film and D_m^* is the molecular diffusivity. D^* is used to distinguish the diffusivity from the deficit D. The rate of mass transport (q) is sometimes written as dm/dt. Letting

$$K_L = \frac{D_m^*}{\delta} \quad (7)$$

Eq. 6 may be written as

$$q = AK_L (c_s - C) = AK_L D \quad (8)$$

K_L is sometimes called the liquid film coefficient or the transfer coefficient. This latter terminology is probably better since the definition of K_L in Eq. 8 is a general definition and not restricted to the concept of a liquid film. The rate of change of the concentration C below the film is related to q by dC/dt = q/V, where V is the volume of water. Thus, Eq. 8 may also be written as

$$\frac{dC}{dt} = K_L \frac{A}{V}(c_s - C) = (K_L a)\ (c_s - C) \quad (9)$$

where a = A/V. The quantity ($K_L a$) is often treated as one coefficient, particularly in case of bubble or spray reaeration because of the difficulty in determining the surface area. Comparison of Eqs. 2, 8 and 9 shows that

$$K_2 = K_L a = \frac{K_L}{h} \quad (10)$$

if h is defined as a mean depth equal to V/A for surface reaeration.

Objections have been raised to at least two assumptions in Lewis and Whitman's film theory. One of these assumptions was the stagnant film at the free surface of a turbulent liquid. Various authors have observed that it was not physically reasonable to make such an assumption. Review of the paper by Lewis and Whitman (1924) does not make it clear whether they considered the film to be physically stagnant. In referring to the film, they used phrases such as "a layer in which motion by convection is slight compared to that in the main body" and "assuming the existence of stationary films." It seems that the essence of their model was not a physically stagnant film, but rather it was a region in which molecular diffusion is the controlling factor in the transport process.

The second assumption which has been questioned was that of a linear dissolved oxygen concentration distribution through the film. The linear distribution strictly corresponds only to a steady state distribution, whereas the concentration beneath the film is continually changing in many gas absorption problems. However, Dobbins (1956) has shown that the linear distribution approximation is a reasonable assumption for oxygen absorption from bubbles. The same conclusion can be reached for most cases of natural reaeration across a free surface by investigation of the unsteady state diffusion equation (Holley et al., 1970).

2.2 Renewal (Statistical) Models

As an alternative to the film theory model, Higbie (1935), Danckwerts (1951), and Dobbins (1956, 1962, 1964) have presented models which are in some ways related to each other. Each of these models sought to account for the effect of mixing on the transport rate away from the free surface, and each used some parameter which basically sought to represent the statistical nature of water being brought to the free surface by turbulence.

Higbie's model for the rate of gas absorption was based on the assumption that the whole body of water was stagnant for short periods of time and, during these periods, oxygen was absorbed and diffused downward solely by molecular diffusion. Then periodically, the water was instantaneously and completely mixed. This model led to an equation which was effectively the same as Eq. 8 with K_L given by

$$K_L = 2\sqrt{\frac{D_m^*}{\pi t'}} \tag{11}$$

where t' is the average time between the complete mixings. Higbie's development has been summarized by Pasveer (1953). This model has been referred to as the penetration model.

Danckwerts (1951) extended Higbie's approach by assuming that various vertical elements of the water could individually undergo complete vertical mixing with different periods between mixings. He assumed that the statistical distribution of the mixing was described by

$$f(t) = re^{-rt} \tag{12}$$

where f(t) is the proportional part of the vertical elements of water for which the elapsed time since the last mixing is between t and t + dt. The constant r may be interpreted as the average rate at which vertical mixing takes place. Thus, r is analogous to 1/t' of Eq. 11. Danckwerts' approach again led to Eq. 8 for the rate of transfer if K_L is given by

$$K_L = \sqrt{D_m^* r} \tag{13}$$

Kishinevsky (1954) raised several objections to Danckwerts' model. One of his primary objections concerned an assumption used by Danckwerts in obtaining Eq. 12, namely the assumption that the probability that an element of water would be mixed was independent of the elapsed time since the last mixing for that element. Other detailed objections were also presented.

Dobbins (1956, 1962, 1964) used the concept of a surface film but postulated that the water in the film was periodically mixed with water from below the film, i.e. new surface film was periodically created and then removed. It was assumed that only molecular diffusion took place while a given parcel of water was in the film. The region below the film was assumed to be uniformly mixed at all times. In a sense, this was a combination of film theory and Danckwerts' work. Dobbins' model is different from Danckwerts' model in that it places a limit (the film thickness) on the depth to which the dissolved gas may move by molecular diffusion. Dobbins also assumed that various parts of the film were mixed or "renewed" at a rate such that Eq. 12 described the distribution of times (or ages) since the last renewal within the film. For a film of thickness δ, it was shown (Dobbins, 1956) that these assumptions lead to the following expression for the transfer coefficient, K_L:

$$K_L = \sqrt{D_m^* r} \coth \sqrt{\frac{r\delta^2}{D_m^*}} \tag{14}$$

As the average surface renewal rate r approaches zero, K_L in Eq. 14 approaches that given by Eq. 7, while for large r, Eq. 14 approaches Eq. 13.

O'Connor and Dobbins (1958) and Dobbins (1964) presented various arguments and relationships for relating r and δ to flow and fluid characteristics in order to arrive at expressions which could be used for predicting the value of K_2 or K_L for field conditions. By various additional assumptions, O'Connor (1958) revised Eq. 14 to read

$$K_L = \sqrt{\frac{D_m^* U}{h^3}} \qquad (15)$$

where U is the average velocity and h is the average depth.

Ruckenstein (1966) considered the partially analogous problems of transport away from a free surface and from a solid boundary. His developments were based on assumptions which more nearly represented the actual turbulence characteristics than did some of the earlier assumptions of the penetration and renewal models. On the basis of dimensional reasoning, for mixing vessels he developed a relationship between K_L, the turbulent energy dissipation, and the fluid properties including D_m^*. For turbulent films on vertical walls, he developed a similar expression. Most of his results were left as proportionalities and were not carried to the stage of obtaining prediction equations.

All of the models mentioned above indicate a dependence of K_L on D_m^*. Thus, they may be invalid for gas transfer at extremely high mixing rates. Kishinevsky and Serebryansky (1956) presented experimental results indicating the same values of K_L for the absorption of gases with different D_m^* values into liquids in which high mixing rates prevailed. For those tests, K_L apparently was independent of D_m^*.

2.3 Empirical Equations

There have been several investigators who used either wholly empirical or partially empirical approaches in attempting to relate reaeration coefficients to mean flow parameters (e.g. Streeter and Phelps (1925), O'Connor and Dobbins (1958), Churchill et al. (1962), Krenkel and Orlob (1963), Thackston and Krenkel (1969), Lau (1972A, 1975B), plus several others cited in the critical reviews mentioned in Section 2.4). Only a few of these approaches will be summarized here.

Churchill et al. (1962) made field measurements of reaeration rates in five streams immediately below water impoundments. The oxygen deficit existed since the released water came from the lower part of the reservoirs where little oxygen was present. There was essentially no BOD present. For these measurements, there were relatively few interferences with the reaeration process. Churchill et al. took a totally empirical approach and tried a large number of possible correlations between k_2 and flow parameters. They concluded by recommending the equation

$$k_2(\text{days}^{-1}) = 5 \frac{U(\text{fps})}{[h(\text{ft})]^{5/3}} \tag{16}$$

where U is the average velocity and h is the average depth.

Thackston and Krenkel (1969) postulated that K_2 should be proportional to the surface renewal rate, which in turn should be proportional to the vertical eddy diffusion coefficient divided by the average depth squared. From both laboratory and field data, Thackston and Krenkel obtained an expression which can be written as

$$K_L = 0.00029(1 + Fr^{0.5})u_{*o} \tag{17}$$

where u_{*o} is the shear velocity and Fr is the Froude number, which was introduced to account for the increase in surface area due to the formation of waves. Harleman and Holley (1963) were apparently among the first to suggest the dependence of reaeration rate coefficients on the vertical eddy diffusion coefficient. This general type of dependence has been used as the basis for semi-empirical models by a number of investigators.

Lau (1972A, 1975B) applied dimensional analysis to the dependence of k_2 on fluid properties and flow parameters. He concluded that some of the prediction equations which have appeared in the literature have not included enough independent variables to completely describe the hydraulics of the flow. He (Lau, 1975) performed laboratory experiments in which he could independently vary the Reynolds number (Re) and the ratio of shear velocity to mean velocity (u_{*o}/U). He concluded that the parameter k_2h/U depends on Re and u_{*o}/U in essentially the same way that the Darcy-Weisbach friction factor (f) depends on Re and the effective relative roughness.

2.4 Critical Evaluations

There have been a number of critical evaluations of various prediction equations which have been proposed for reaeration rate coefficients. See, for example, the evaluations of Bennett and Rathbun (1972), Lau (1972B), Kramer (1974), Brown (1974), Wilson and Macleod (1974), and Rathbun (1977). The different authors use various methods of discussing and evaluating the prediction equations. One common method is to compare calculated values of K_2 for various equations with experimental values from various publications. Bennett and Rathbun (1972) made a direct comparison of k_2 values for assumed hydraulic conditions using 13 prediction equations from the literature. The result of their comparison is shown in Fig. 2. The curves labelled N and O were added by

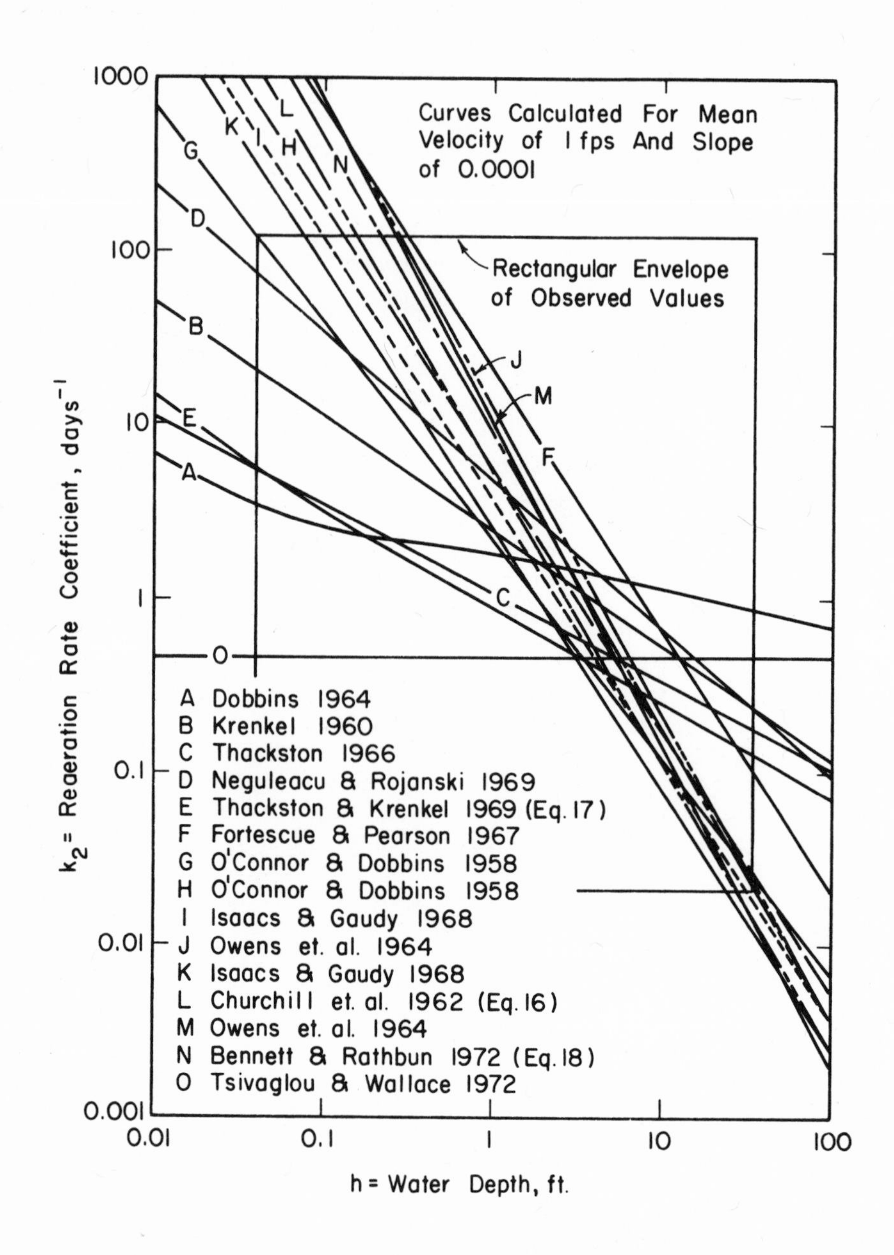

Fig. 2. Comparison of Reaeration Rate Coefficients from Various Prediction Equations (after Bennett and Rathbun, 1972).

the writer. The references from which the prediction equations were obtained are shown on the figure. The various lines in the figure represent calculated values of k_2 as a function of depth for a constant velocity and slope. One could question the validity of this type of comparison since it represents an extrapolation of empirical results. Also, using Manning's equation, it is possible to see that some of the assumed conditions are not physically realizable within a realistic range of Manning's n values, and some of the depths are not very realistic for rivers and streams. Nevertheless, Fig. 2 serves to illustrate the fact that large differences can exist between values of k_2 calculated from various equations in the literature.

Most investigators have sought to develop prediction equations based on both field and laboratory data. Bennett and Rathbun (1972) took a different approach and separated the available data into field data and laboratory data. They performed various regression analyses for these separate sets of data to obtain prediction equations for k_2. They concluded that all the resulting equations obtained from laboratory data were significantly different statistically from those obtained from field data. Curve N in Fig. 2 represents the empirical equation which Bennett and Rathbun (1972) obtained as the best fit to all of the available field data, namely

$$k_2(\text{days}^{-1}) = 8.76 \, \frac{[U(\text{fps})]^{0.607}}{[h(\text{ft})]^{1.689}} \tag{18}$$

Most of the work on reaeration prediction equations has related to situations where the flow was taking place in well-defined channels with the depth of flow being much greater than the height of irregularities or roughness elements on the channel boundaries. These hydraulic conditions do not always exist in natural streams, especially under low flow conditions. Even though there are significant water quality problems associated with low flow conditions, there apparently have been relatively few published reports of reaeration studies under low flow condition. One such study was made by Foree (1975) for streams in Kentucky. He obtained an empirical relation between k_2 and stream slope.

Rathbun (1977) cites references which consider the prediction equations as a function of discharge, and he points out that the K_2 values predicted by several of the equations tend to converge at higher discharges but to diverge at lower discharges. This conclusion and the conclusions of Bennett and Rathbun (1972) point to a definite need to give more consideration to the conditions under which a prediction equation was obtained and not to apply an equation out of its range of applicability. The various evaluations of prediction equations illustrate that the understanding of the reaeration process and the development of prediction equations for rate coefficients such as K_2 or K_L are certainly not complete. The following three sections discuss some concepts which may prove helpful in further studies of the reaeration process.

3. BOUNDARY LAYER CONCEPTS

Many transport problems in turbulent flow can be treated by a diffusion law analogous to Fick's law for molecular diffusion (Bird et al, 1960), and it should be possible to use a diffusion model to represent the transport of oxygen downward from the free surface in turbulent water. By analogy to Fick's law, a general transport equation for diffusion in the vertical (y) direction may be written as (Bird et al, 1960)

$$q = -AD^* \frac{\partial c}{\partial y} \tag{19}$$

where D^* is a diffusion coefficient and c is a temporal mean concentration in which turbulent fluctuations have been averaged out. In general, D^* may be written as $D_m^* + D_t^*$, where the subscript m refers to the molecular contribution to diffusion and the subscript t refers to the turbulent contribution, and the numerical value of D^* depends on the particular mixing conditions.

The use of Eq. 19 to represent the transport away from the free surface can be referred to as a "diffusion model." Harleman and Holley (1963) and Thackston and Krenkel (1969), among others, have discussed the general significance of vertical diffusion in the reaeration problem, but such publications have not given detailed consideration to the transport immediately below the free surface. Ruckenstein (1966) presented some considerations relating to the possible variation of D^* below a free surface in turbulent water.

The actual physical conditions occurring during reaeration, of course, are not affected by the analytical model which is used. The assumption that the water surface is at saturation concentration is as valid for a diffusion model as for any other model. Also, for unstratified turbulent flow it can be assumed as stated earlier that the water at some distance below the surface is well enough mixed so that the oxygen concentration is essentially uniform. In view of these assumptions, the time-averaged concentration distribution must be essentially as shown in Fig. 1B. That is, in a diffusion model there would be a relatively small region immediately below the surface with a steep gradient from saturation concentration at the surface to the concentration existing in the bulk of the water. A diffusion model (Eq. 19) could be used to represent the transport downward from the free surface in terms of this gradient immediately below the free surface.

The region containing the steep gradient near the surface is normally called a surface film or a liquid film, but it is actually an "oxygen boundary layer" and it would be reasonable to refer to it as such. This term would call attention to the analogy between oxygen boundary layers and other boundary layers. This analogy and the general behavior of such boundary layers has been overlooked in the development of some of the prediction equations for reaeration rate coefficients. Due to past successes in correlating diffusion coefficients and boundary layer thickness with mean flow parameters for various transport processes, there exists the possibility of finding such correlations for the reaeration phenomenon. However, a problem arises due to the fact that the oxygen boundary layer is so thin that with currently available instruments and experimental techniques the details of the oxygen distribution within the boundary layer can not be measured. Therefore, neither the oxygen boundary layer thickness nor the diffusion coefficient in the boundary layer can be determined directly. It is necessary to try to infer some things about the oxygen transport process from what is known about other types of boundary layers.

Holley (1973) considered some fundamental boundary layer concepts and some data on mass transfer boundary layers to draw several conclusions about the expected behavior of surface films or oxygen boundary layers. Some of his conclusions were the following: The thickness (δ) of the surface film should be a function of both the Schmidt number (Sc) and the mixing conditions in the water, as represented for example by the Reynolds number (Re) and boundary roughness for flowing water; for constant mixing conditions, δ should decrease with increasing Sc; for constant Sc, δ should decrease with increasing Re; it is possible to have mass transport through a boundary layer at a rate determined by the molecular diffusivity even when turbulence is present in the boundary layer; and indirect evidence suggests that the effective diffusion coefficient in Eq. 18 for oxygen transport is actually the molecular diffusivity (D^*_m), especially for low and moderative levels of mixing, and that most of the influence of mixing on reaeration may be in terms of changes in δ rather than changes in D^* within the film or boundary layer. Lau (1975A) concluded that his experimental results on reaeration rates were compatible with the boundary layer and diffusion concepts of the transport in the film as presented by Holley (1973).

The concept of the film as a boundary layer supports the possibility that θ (Eq. 5) is a function of mixing conditions. Also, this concept would indicate that care should be used in comparing absorption rates for different gases, even under the same mixing conditions, since the values of K_L should depend on both D^*

and δ which varies with Sc. The conclusions concerning the boundary layer nature of the surface film would suggest that for many mixing conditions, efforts to develop analytical models to represent reaeration should focus attention on the dependence of δ on hydraulic conditions since D^* may be nearly constant and equal to D_m^*.

4. SUSPENDED PARTICLES

One factor which has not been thoroughly investigated, and which is often not considered in the literature, is the influence of suspended particles on reaeration rate coefficients. Some information on this influence is summarized in this section.

There are some apparent inconsistencies in the available data on the influence of suspended solids on absorption rates. Kehr (1938) noted that addition of diatomaceous earth to tap water reduced the transfer rate by about 25%. Holdroyd and Parker (1952) reported that bentonite had no effect on the transfer rate. Poon and Campbell (1967) found that suspended particles in small concentrations enhanced the transfer process. Van der Kroon (1968) studied the effects of suspensions of clay and aluminum sulfate. His results seem to indicate a reduction in transfer rate with an increasing suspension concentration. It is not clear whether the changes which he observed in the transfer rate were due to changes in the saturation concentration or changes in $K_L a$ (or both). Eckenfelder et al (1956) found that the addition of 2260 ppm of suspended solids to an activated sludge waste mixture caused α to decrease from 0.54 to 0.44, where α is the ratio of reaeration rate coefficient with a suspension or waste material to the coefficient for plain water. At least part of the apparent inconsistencies are no doubt due to the fact that various types of suspended materials affect the reaeration process in different ways, especially if the suspension results from sewage or industrial wastes. Many of the references did not distinguish between the various mechanisms which might have been affecting the reaeration. Depending on the chemical and physical nature of the suspension and the associated materials present in the water, there are at least four ways in which the reaeration process may be affected, namely, by changes in c_s, by the presence of surface active agents, by the creation of oily films on the surface, and by changes in the turbulence structure of the water. The remainder of this section is concerned with the effect of the physical presence of suspensions on K_L because of changes in the turbulence structure. Also the range of concentration of suspended material to be considered is related to the range to be expected in streams. Much higher concentrations and different types of effects may be encountered in waste treatment facilities (Baker et al, 1975).

Holley et al.(1970) performed mixing-vessel experiments with plain water at 20°C and at 2°C and with water with micro-organisms at the same two temperatures. See also Micka et al. (1973). There were no bubbles created by the mixing and the water surface was approximately horizontal. The purpose of the 2°C tests was to allow the evaluation of the effects of the physical presence of the micro-organisms in the absence of biological activity. The experimental results indicated that there was no oxygen demand for the 2°C tests. There were only three tests with micro-organisms at each temperature, giving the results shown in Fig. 3. The parameter α in the figure is defined as the ratio of K_L with the micro-organisms to K_L for plain water. The data is rather limited, but it shows approximately the same value of α for each temperature. Thus, for these tests, it was concluded that the physical presence of the micro-organisms caused K_L to be increased about 50% over the corresponding value for plain water for the same average hydraulic conditions in the mixing vessel. The concentrations of suspended solids in these tests were in the range of 125 to 300 mg/l.

Tsao (1968) also studied the effect of several types of bacterial cells on the oxygen transfer rate. The experiments were performed in a beaker at low stirrer speeds so that no bubbles were formed. He found values for α on the order of 1.4 when there was biological oxidation during the oxygen transfer process. He assumed that the increase was due to the direct absorption of oxygen from the atmosphere by the cells. The experiments discussed above (Holley et al., 1970; Micka et al., 1973) showed an increase in reaeration rate in a situation where any direct uptake by the cells was not included in the measurements.

Tsivoglou and Wallace (1972) conducted reaeration experiments in a mixing vessel with water containing a suspension of droplets of mineral oil. For constant mixing conditions, they found that α increased from 1.0 to about 2.3 as the concentration of oil increased from zero to about 450 mg/l, as shown in Fig. 4. For a fixed concentration of 230 mg/l, α decreased as the intensity of mixing was increased. The significance of this is not clear. The authors did not comment on any observed changes when the stirrer speed was increased. The oil droplet size probably was reduced by the increased mixing and this change may have accounted for the changes in K_L.

Alonso et al. (1975) studied the effects of suspensions on reaeration rates in open channel flows in the laboratory. In the experiments, uniform flow was established and then various amounts of 0.115-mm sand were added and k_2 was determined for different concentrations of suspended sand. The results were scattered, but indicated about a 35% decrease in dimensionless k_2 values as the concentration of suspended sand increased up to about 3500 ppm.

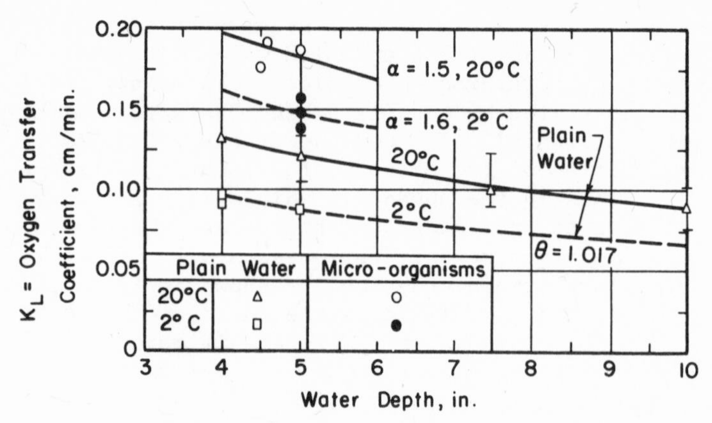

Fig. 3. Effects of Suspensions of Micro-organisms on Oxygen Transfer Coefficient for a Mixing Vessel (after Holley et al., 1970, and Micka et al., 1973).

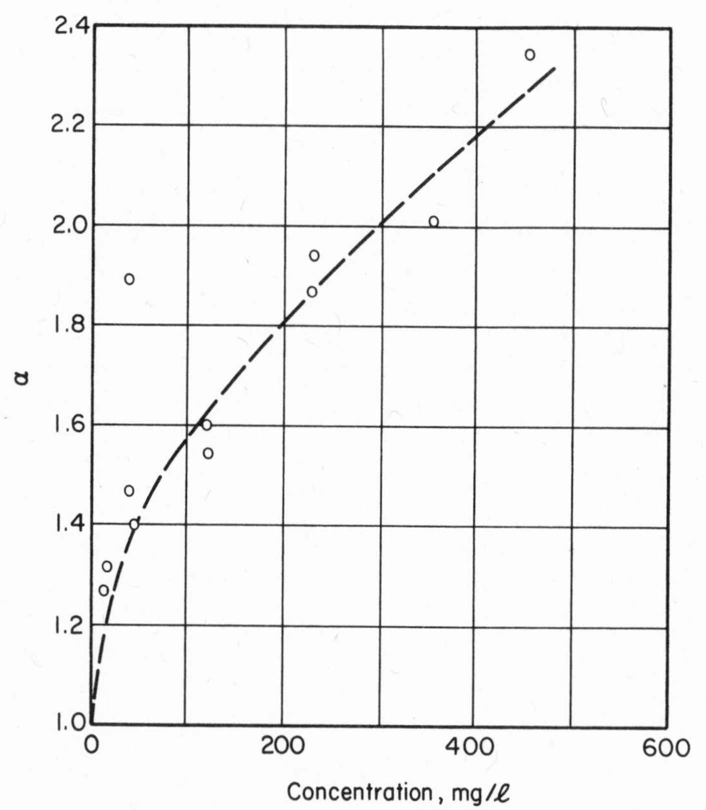

Fig. 4. Effects of Suspensions of Oil Droplets on Oxygen Transfer Coefficient for a Mixing Vessel (after Tsivoglou and Wallace, 1972).

The reason for this decrease in k_2 in view of the previously cited increases in reaeration rate coefficients is not clear, but there is one possible explanation which emphasizes a point which must be considered in studying the effects of suspensions. In correlating their results, Alonso et al. used a dimensionless group equal to k_2R/u_{*o} where R is the hydraulic radius and u_{*o} is the bed shear velocity. Also, they accounted for decreases in von Karman's k with increasing sediment concentrations. However, from the description of their experiments and data analysis, it is not clear whether they accounted for changes in u_{*o} with changes in sediment concentration. If they did not, then it should be expected that as the sediment concentration increased for a given depth and velocity, the resistance to flow and the value of u_{*o} decreased (Vanoni, 1946). This, in turn, would mean that for a given depth and velocity, the energy dissipation decreased and less energy was going into turbulent mixing as the sediment concentration increased. This type of behavior could explain the observed decreases in k_2R/u_{*o} since reaeration rate coefficients are generally proportional to the vertical turbulent diffusion coefficient (Section 2.3). In the various mixing vessel experiments cited earlier, the changes in reaeration coefficients were based on situations with constant depths and approximately constant energy inputs to the mixing. In comparing reaeration rate coefficients to study the effects of suspended sediments, it is important to consider possible changes in the energy input to the flow and to the mixing process.

Thus, it can be concluded that there is evidence that the presence of suspended substances can significantly change the value of K_L (or equivalently k_2) and most of the evidence points to an increase in K_L due to the physical presence of suspensions. Nevertheless, the concentration of suspensions has not generally been considered as a significant variable in seeking correlations between reaeration rate coefficients and flow and fluid parameters. An increase in K_L is consistent with what is known about the general dependence of K_L on turbulent mixing and about the effects of suspended solids on turbulent mixing. It is well documented that the presence of suspended material in turbulent flow can cause a significant increase in the turbulent diffusion coefficient, D_t^*. This can be verified from the experiments of Vanoni (1946), Ismail (1952), and Apmann and Rumer (1970), for example. Householder and Goldschmidt (1969) summarized a large amount of data from the literature and indicated increases in turbulent diffusivities for open channels up to a factor of three due to suspended particles. This type of increase is generally attributed to the fact that the suspended material causes a redistribution of the turbulent energy so that more of the energy occurs in the lower frequency components. On the basis of the dependence of K_L or K_2 on turbulent mixing, this data indicates that an increase in reaeration rate coefficients should be expected when suspended material is present and that suspended sediments should be considered as one of the variables which can influence the magnitude of reaeration coefficients.

5. WIND

5.1 General Effects of Wind

Much of the preceding discussion has been directed primarily toward reaeration or oxygen absorption in flowing streams where the turbulence is a result of the shear at the channel boundaries. Nevertheless, the basic concepts and mechanisms are the same regardless of the origin of the turbulence and would be equally applicable to turbulence and mixing generated by wind blowing over a water surface. Wind is especially effective in the reaeration process in either flowing or non-flowing water since the wind shear is applied to the free surface of the water and thus creates mixing in the surface film region where the primary resistance in the oxygen adsorption process is located. Neglecting the influence of wind may have caused a significant part of the deviations between certainly one of the primary causes of reaeration in large bodies certainly one of the primary causes of reaeration in large bodies of water where the flow velocities are small.

There has been a recognition of the effects of wind on reaeration in natural water bodies at least since the often-quoted work of Downing and Truesdale (1955). In spite of the time that has elapsed since their work, there have been relatively few additional studies on wind effects. From the studies which have been made, a general and somewhat obvious conclusion can be drawn that the dependence of the reaeration process (as represented, for example, by the magnitude of K_L) depends on both the wind speed (U_a) and the fetch of the wind. For U_a less than some critical value there is apparently a negligible effect of wind on K_L. The absence of wind effects is demonstrated by the glassy-smooth water surface that can be observed under calm or nearly calm conditions for some bodies of water. Wind speeds of about 5 to 7 m/sec at 10 meter height are normally quoted as the critical value above which the wind begins to influence the reaeration process (Kanwisher, 1963A,B; Banks, 1975). For U_a greater than the critical value, there is mixed evidence on whether K_L increases as U_a to the first power (Thames Survey Committee, 1964, as quoted by Banks, 1975) or to some power greater than unity (Downing and Truesdale, 1955). Powers as great as two have been suggested for the higher velocity range (Kanwisher, 1963A,B; Banks, 1975). The physical reasoning for suggesting powers greater than unity for the variation of K_L with U_a is usually based on the changes in the water surface as U_a increases. As the wind speed increases following calm conditions, ripples begin to develop. These ripples increase the drag of the wind on the water and thereby increase the energy transfer from the wind to the water. These increases result both in more turbulence, especially near the surface, and in an increased surface area through which absorption can take place. The changes in both the turbulence and the surface area give increased reaeration, with the changes in turbulence accounting for most of the increased reaeration, especially at the lower wind speeds. As the wind speed continues to increase, the ripples progress to waves, then to breaking waves, and then to breaking waves with spray. The breaking waves and spray give significant increases in surface area as droplets are formed. It is reasoned (Banks, 1975) that these progressively greater disturbances of the water surface could produce progressively greater rates of increase of K_L relative to U_a. These changes in the water surface conditions have been discussed by Kanwisher (1963A,B) and by Banks (1975). Kanwisher (1963A) shows photographs of sea surface conditions for various wind speeds.

There has also been some reference (Eloubaidy and Plate, 1972) to the fact that wind enhances evaporation which tends to cool the water surface and thereby to increase the saturation concentration (c_s) at the surface. While such an increase in c_s would increase the oxygen absorption rate (Eq. 2) it would not increase K_L (Eq. 8) unless the cooling became sufficient to cause a density instability at the surface and to thereby enhance the mixing at the surface.

5.2 Laboratory Data

Downing and Truesdale (1955) studied the influence of several variables, including wind, on reaeration. Their experiments on wind effects were done in a mixing tank 0.92 m long, 0.30 m wide, and 0.38 m deep. Air was blown horizontally over the water surface and low intensity mechanical stirring was provided in the tank. Some other conditions related to their experiments and to other experiments to be discussed below are given in Table I. Downing and Truesdale's laboratory values of K_L were converted by the writer from 15°C to 20°C by assuming $\theta = 1.024$ (Eq. 5). The data are shown in Fig. 5.

Kanwisher (1964A,B) conducted experiments to determine surface film coefficients for the absorption of carbon dioxide. The experiments were done in a tank 1 m long, 0.5 m deep, and 0.5 m wide in a wind tunnel. Since the molecular diffusivity for carbon dioxide in water is approximately equal to that for oxygen in water (Washburn, 1929), Kanwisher's K_L values should be approximately the same as those for oxygen. He reasoned that the reaction of carbon dioxide with water did not significantly affect the absorption process in his experiments.

Eloubaidy and Plate (1972) conducted reaeration experiments in a wind-water tunnel 15.8 m long and 0.61 m wide with a water depth of 0.12 m. They reported increases of up to 20% in K_L with increasing fetch. They did not publish the individual measurements but presented a semi-empirical equation which had been fitted to

Table I. Experimental Conditions for Studies of the Effects of Wind Speed on K_L

Source	Lab./ Field	Symbol in Fig. 5	Water Depth	Water Velocity	Height of Measured Wind Speed	Wind Speed	NaCl Concen- tration	Mechanically Generated Waves	Temp.
			m	m/sec	m	m/s	%		°C
Downing, Truesdale 1955	L	1	0.25	stirred	0.05	0.9-13.4	3	Yes	15
	L	2	0.25	stirred	0.05	0-6.3	0	Yes	15
	L	3	0.25	stirred	0.05	0-8.9	0	No	15
	F	○	≈5(?)	>0	10	0,10	?	-	?
Kanwisher 1963A,B	L	4	0.50	0	0.10	0.1-9.8	0	Yes	?
	L	5	0.50	0	0.10	0.1-10	0	No	?
Thames Survey 1964	F	———	≈5(?)	>0	10		?	-	?
Juliano 1969	F	+	7.3	>0	?	0-2.7	>0	-	?
	F	×	7.6	>0	?	0.9	>0	-	?
	F	◇	7.6	>0	?	0	≈0	-	?
	F	△	2.1	>0	?	3.6-5.4	>0	-	?
	F	□	4.9	>0	?	0-6.7	>0	-	?
Eloubaidy, Plate 1972	L		0.12	0.23-0.36	0.6	6.7-11.6	0	No	?
Kramer 1974	F	●	11.6	>0	?	2.2	≈3	-	?
Banks 1975	L	———	-	-	10	-	0	-	15
Mattingly 1977	L	6	0.27	0.18	0.20	0-15	0	No	20
	L	7	0.27	0.09	0.20	0-15	0	No	20
	L	8	0.27	0.045	0.20	0-15	0	No	20

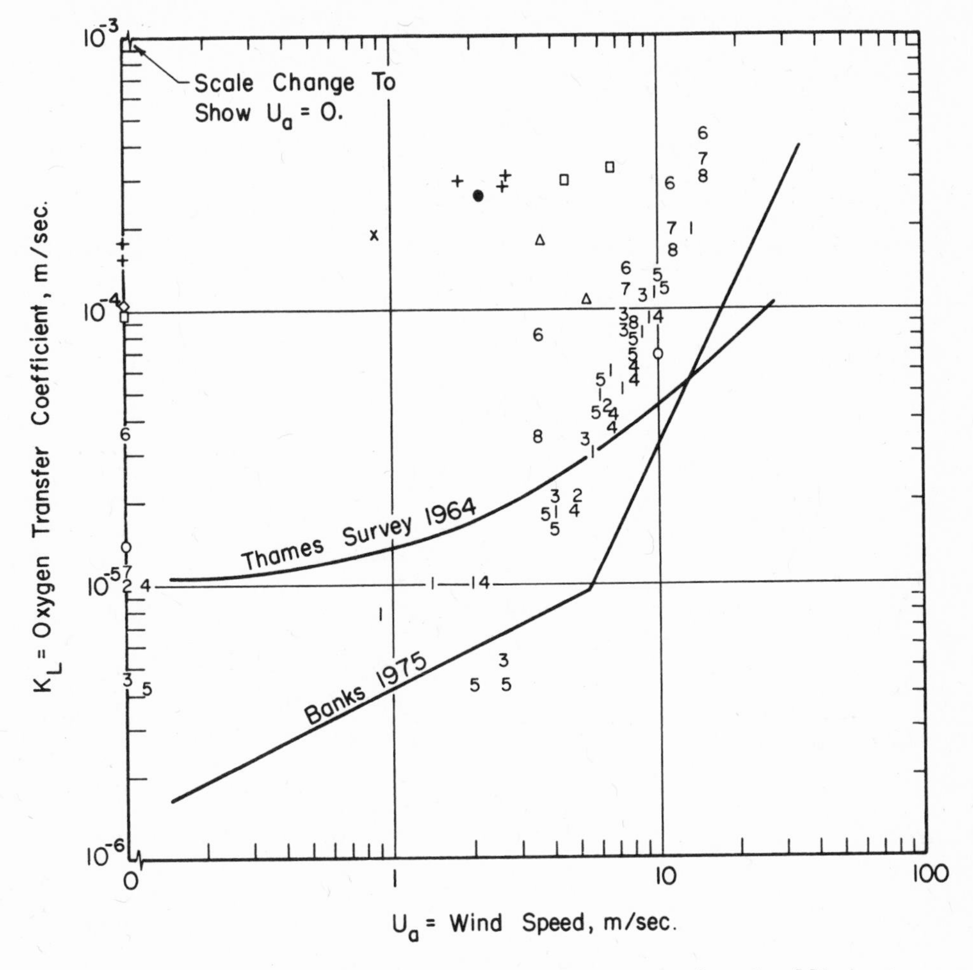

Fig. 5. Effect of Wind Speed on Oxygen Transfer Coefficient.

the K_L values for the last 2.44-m long segment of the flume. Their equation can be written as

$$K_L = 7.20 \times 10^{-8} \; u_{*t} \; \frac{u_{*s}h}{\nu} \tag{20}$$

where, for two-dimensional flow, $u_{*t} = \sqrt{(\tau_o - \tau_s)/\rho}$, $u_{*s} = \sqrt{\tau_s/\rho_a}$, τ_o = boundary shear stress, τ_s = water surface shear stress, ρ = density of water, ρ_a = density of air, h = depth of water, and ν = kinematic viscosity of water. Eloubaidy and Plate stated that Eq. 20 is valid only for U_a > 2 m/sec, which would correspond to approximately 4 m/sec at 10 m height. They suggested using

$$u_{*s} = 0.0184\ U_a^{3/2} \tag{21}$$

where u_{*s} and U_a are in m/sec and U_a is the velocity "outside of the boundary layer along the water surface". The definitions given by Eloubaidy and Plate (1972) appear to be based on standard sign convention, which means that τ_o and τ_s are both positive when the wind is blowing in the direction of the water flow. For Eloubaidy and Plate's experiments, τ_s was in the range of about 10% to 50% of τ_o. From the one-dimensional momentum equation, it can be seen that it is not possible to have $\tau_s \geq \tau_o$ for uniform flow in a wind-water tunnel with a positive bottom slope. However, for field situations it is possible to have $\tau_s \geq \tau_o$ and then Eq. 20 is without meaning since u_{*t} becomes imaginary. Bennett and Rathbun (1972) indicated that Eq. 20 was verified with very limited data. Rathbun (1972), in a discussion of Eloubaidy and Plate's paper, compared Eq. 20 with data from Churchill et al (1962), O'Connor and Dobbins (1958), and Owens et al. (1964). The ratio of predicted to observed k_2 values showed a steady increase with increasing water depth. The ratio increased from 0.4 at 0.4 ft depth to 70 at 20 ft depth. Rathbun (1972) suggested that Eq. 20 should not be used outside the range of depths for which it was verified in the laboratory. Because of these various difficulties with Eq. 20, it has not been shown in Fig. 5. Nevertheless, Eq. 20 is apparently an indication that Eloubaidy and Plate's data showed a dependence of K_L on the hydraulic conditions of the channel flow as well as on U_a and τ_s.

Banks (1975), based on the changing nature of the interaction of wind and the water surface as cited above, obtained the following three semi-empirical equations for different ranges of U_a:

$$K_L = 4.19 \times 10^{-6} U_a^{1/2}, \quad \text{small } U_a \tag{22}$$

$$K_L = 1.80 \times 10^{-6} U_a, \quad \text{intermediate } U_a \tag{23}$$

$$K_L = 0.32 \times 10^{-6} U_a^{2}, \quad \text{large } U_a \tag{24}$$

where K_L and U_a are in m/sec and U_a is measured at a 10 m height. Banks obtained the numerical coefficients in these three equations by matching the equations to the laboratory data of Downing and Truesdale (1955) with the wind speeds converted to their equivalent 10 meter values. The intermediate classification can effectively be dropped since Banks' results show that the intermediate range is negligibly small. It should be noted that the curves in Fig. 5 for Eqs. 22 and 24 are based on wind speeds at 10-m height while the laboratory data points are plotted on the basis of the values of U_a actually measured in the experiments.

Banks (1975) compared K_L values to heat transfer coefficients for water surfaces. While reaeration and heat transfer (at least the parts of heat transfer associated with evaporation and diffusion (or conduction) in the air) are definitely both related to the wind speed and fetch, it is not clear that there should be numerical equality of the analogous coefficients since K_L depends on the wind-induced mixing in the water while evaporation and diffusion in the air depend on wind-induced mixing in the air. Furthermore, the expression which Banks used for evaluating the heat transfer coefficients included some radiation transfer which has no analogy in the gas transfer problem.

Mattingly (1977) performed reaeration experiments in a laboratory wind-water tunnel 0.55 m wide with water and air depths of 27 cm. The test section was 8 m long. In addition to the reaeration, he also measured velocity distributions in the air and obtained some qualitative information on the turbulent velocity fluctuations in both the air and the water. His results showed a slight but a definite dependence of K_L on water velocity in addition to the primary dependence on U_a even though the water velocities were an order of magnitude smaller than U_a.

There is remarkably good agreement between the three sets of laboratory data shown in Fig. 5, especially for $U_a > 3$ m/sec which corresponds to approximately $U_a > 6$ m/sec at 10 m height. For $U_a < 3$ m/sec, K_L should depend primarily on the water flow rather than on U_a so a large scatter in K_L vs. U_a should be expected. The agreement for $U_a > 3$ m/sec is somewhat surprising in view of the different sizes of the experimental equipment, the different experimental procedures, and documented dependence of K_L on the fetch of the wind and on the height and frequency of surface waves (Downing and Truesdale, 1955; Eloubaidy and Plate, 1972; Mattingly, 1977). Although it was not stated explicitly by all of the authors, the wind was apparently in the same direction as the water flow for the laboratory experiments with water flow.

5.3 Field Data

The field data on wind effects on reaeration is extremely limited and inconclusive. Some of the conditions for the data which could be located by the author are given in Table I and the data is plotted in Fig. 5 together with the laboratory data. Downing and Truesdale (1955) make reference to field studies in the Thames Estuary and give enough information for two measurements to allow those two points to be shown on Fig. 5. Banks (1975) cited the equation

$$K_L = (10.0 + 3.38U_a) \times 10^{-6} \tag{25}$$

from the Thames Survey Committee (1964), based on field measurements in the Thames Estuary. In Eq. 25 K_L and U_a are in m/sec. Eq. 25 gives $K_L \sim U_a$ while the laboratory data indicate $K_L \sim U_a^n$ with $n \simeq 2$ for the higher velocities.

Juliano (1969) conducted reaeration experiments in various parts of the Sacramento-San Joaquin estuary. Some experiments were done using the gasometric method with a floating dome and some were done with the disturbed equilibrium method by using sodium sulfite to deoxygenate the water in a submerged, floating 15-ft diameter plastic swimming pool with the bottom removed. This latter procedure naturally caused some disturbance of the natural mixing patterns for the water for which the reaeration was being measured. The values of K_L and U_a which could be determined from the published data are shown in Table I and on Fig. 5.

Kramer (1974) quoted a single value of K_2 for the Houston Ship Channel, based on the work of Hann et al (1972). This value of K_2 and typical data given by Kramer for the Ship Channel were used to obtain the K_L-U_a value shown in Fig. 5.

5.4 Conclusions on Effects of Wind Speed

There are two tentative conclusions that can be drawn from Fig. 5. One conclusion is that in spite of the good agreement among the various sets of laboratory data, there are significant differences between the laboratory data and the limited field data both in terms of the values of K_L over most of the range of U_a values and in terms of the trend of the changes in K_L with increasing U_a. A second conclusion is that there is approximately an order of magnitude difference between Eq. 25 obtained from the Thames data and most of the rest of the field data, but the general trend of variations in K_L vs. U_a for Eq. 25 appears similar to that for the remainder of the field data. It may be that physical-chemical interferences with the reaeration process in the Thames Estuary were responsible for the differences between Eq. 25 and the remainder of the field data. It should be noticed that most of the laboratory data was obtained for values of U_a large enough that the primary dependence of K_L might be expected to be on U_a. It is not immediately apparent that the same dependence would be true for the field data since most of it was obtained for smaller U_a values.

Based on the available data, there would seem to be a need for further research into the effects of wind on reaeration to investigate the differences between laboratory and field data, to obtain more definitive field data on the dependence of K_L on U_a, to study the significance of relative directions and magnitudes of wind and water flow, and to further evaluate the significance of wind fetch and natural wave characteristics.

6. SUMMARY

A review has been given of some basic concepts related to the reaeration process and of some of the analytical models and prediction equations which have been proposed for reaeration rate coefficients in streams. There is no general agreement between the various models and equations. A completely satisfactory understanding and representation of the reaeration process is yet to be achieved.

Some considerations are presented which may be helpful in formulating future studies of reaeration. One thing to be considered is that the surface film is actually a mass transfer boundary layer. Boundary-layer or film thicknesses are a function of both mixing conditions and Schmidt number. Also, it is possible to have transport in the boundary layer taking place at a rate given by the molecular diffusivity even when turbulence is present in the boundary layer. Another factor is that suspended particles in the flow can have a significant influence on the turbulence structure and therefore on the reaeration rate coefficients, as compared to the coefficients for flow of plain water at the same mean flow parameters. Both laboratory and field data indicate a dependence of K_L on wind speed for both flowing water and large water bodies where the flow velocities are negligible. However, there are significant differences between the laboratory and the field data on the effects of wind speed.

There are many aspects of the reaeration process which need further study.

7. REFERENCES

Alonso, C. V., J. R. McHenry, and J.-C.S. Hong, "The Influence of Suspended Sediment on the Reaeration of Uniform Streams," Water Research, vol. 9, pp. 695-700, 1975.

ASCE Committee on Sanitary Engineering Research, "Solubility of Atmospheric Oxygen in Water," J. San. Engr. Div., ASCE, vol. 86, no. SA4, pp. 41-53, July 1960.

Apmann, R. P. and R. R. Rumer, "Diffusion of Sediment in Developing Flow," J. Hydr. Div., ASCE, vol. 96, no. HY1, pp. 109-123, Jan. 1970.

Baker, D. R., R. C. Loehr, and A. C. Anthonisen, "Oxygen Transfer at High Solids Concentrations," J. Envir. Engr. Div., ASCE, vol. 101, no. EE5, pp. 759-774, Oct. 1975.

Banks, R. B., "Some Features of Wind Action on Shallow Lakes," J. Envir. Engr. Div., ASCE, vol. 101, no. EE5, pp. 813-827, Oct. 1975.

Bennett, J. P. and R. E. Rathbun, "Reaeration in Open-Channel Flow," Prof. Paper 737, U.S. Geological Survey, Gov't Print. Off., Washington, 75 p., 1972.

Bird, R. B., W. E. Stewart, and E. N. Lightfoot, Transport Phenomena, Wiley, New York, 1960.

Brown, L. C., "Statistical Evaluation of Reaeration Prediction Equations," Envir. Engrg. Div., ASCE, vol. 100, no. EE5, pp. 1051-1068, Oct. 1974.

Churchill, M. A., H. L. Elmore, and R. A. Buckingham, "The Prediction of Stream Reaeration Rates," J. San. Engr. Div., ASCE, vol. 88, no. SA4, pp. 1-46, July 1962.

Danckwerts, P. V., "Significance of Liquid Film Coefficients in Gas Absorption," Ind. and Engr. Chem., vol. 43, pp. 1460ff, June 1951.

Dobbins, W. E., "The Nature of the Oxygen Transfer Coefficient in Aeration Systems," Chap. 2-1 in Biological Treatment of Sewage and Industrial Wastes, ed. J. McCabe and W. W. Eckenfelder, vol. 1, Reinhold, New York, 1956.

Dobbins, W. E., "Mechanism of Gas Absorption by Turbulent Fluids," paper presented at Intern. Conf. on Water Pollution Research, London, Sept. 3-7, 1962.

Dobbins, W. E., "BOD and Oxygen Relationships in Streams," J. San. Engr. Div., ASCE, vol. 90, no. SA3, pp. 53-78, June 1964.

Downing, A. L. and G. A. Truesdale, "Some Factors Affecting the Rate of Solution of Oxygen in Water," J. Appl. Chem., vol. 5, pp. 570-581, 1955.

Eckenfelder, W. W., W. R. Lawrence and D. T. Lauria, "Effect of Various Organic Substances on Oxygen Absorption Efficiency," Sew. and Ind. Wastes, vol. 28, no. 11, pp. 1357-1364, 1956.

Edinger, J. E., discussion of "Impoundment and Temperature Effect on Waste Assimilation" by P. A. Krendel, E. L. Thackston, and F. L. Parker, J. San. Engr. Div., vol. 95, no. SA5, pp. 991-994, Oct. 1969.

Eheart, J. W., "Two-Dimensional Water Quality Modeling and Waste Treatment Optimization for Wide, Shallow Rivers," Ph.D Thesis, U. Wisc. Madison, 371 p., 1975.

Eloubaidy, A. F. and E. J. Plate, "Wind Shear-Turbulence and Reaeration Coefficient," J. Hydr. Div., ASCE, vol. 98, no. HY1, pp. 153-170, Jan. 1972.

Engelund, F., "Dispersion of Floating Particles in Uniform Channel Flow," J. Hydr. Div., ASCE, vol. 95, no. HY5, pp. 1149-1162, July 1969.

Fair, G. M. et al, Water and Wastewater Engineering, vol. 2, Wiley, New York, 1968.

Fan, L. T. and S. N. Hong, "Distributed Discharge of Cooling Water Along Direction of Stream Flow," Water Resources Bulletin, vol. 8, no. 5, pp. 1031-1043, Oct. 1972.

Foree, E. G., "Low-Flow Reaeration and Velocity Characteristics of Small Streams," presented at ASCE Hydr. Div. Symp. on Reaeration Research, Gatlenburg, Tenn., Oct., 1975.

Fortescue, G. E. and J.R.A. Pearson, "On Gas Absorption into a Turbulent Liquid," Chem. Eng. Sci., vol. 22, pp. 1163-1176, 1967.

Hann, R. W. et al, "Atmospheric Reoxygenation in the Houston Ship Channel," Estuarine Systems Project Rept 23, Texas A&M Univ., College Station, 1972.

Harleman, D.R.F. and E. R. Holley, discussion of "Turbulent Diffusion and the Reaeration Coefficient," by P. A. Krenkel and G. T. Orlob, Trans., ASCE, vol. 128, part III, pp. 327-333, 1963.

Higbie, R., "The Rate of Absorption of a Pure Gas into a Still Liquid during Short Periods of Exposure," Trans., AIChE, vol. 31, p. 365, 1935.

Holdroyd, A. and H. B. Parker, "Investigation on the Dynamics of Aeration," J. and Proc., Inst. Sew. Purif., pt. 4, pp. 280-305, 1952.

Holley, E. R., "Turbulence Measurements near the Free Surface of an Open Channel Flow," Water Resources Research, vol. 6, no. 3, June 1970.

Holley, E. R., T. Micka, H. Pazwash, and F. W. Sollo, "Effects of Oxygen Demand on Surface Reaeration," Research Report 46, Water Resources Center, University of Illinois at Urbana-Champaign, 80 p., 1970.

Holley, E. R., "Diffusion and Boundary Layer Concepts in Aeration through Liquid Surfaces," Water Research, vol. 7, pp. 559-573, 1973.

Householder, M. K. and V. W. Goldschmidt, "Turbulent Diffusion and Schmidt Number of Particles," J. Engr. Mech. Div., ASCE, vol. 95, no. EM6, pp. 1345-1367, Dec. 1969.

Isaacs, W. P. and A. F. Gaudy, Jr., "Atmospheric Oxygenation in a Simulated Stream," J. San. Engr. Div., ASCE, vol. 94, no. SA2, pp. 319-344, April 1968.

Ismail, H. M., "Turbulent Transfer Mechanism and Suspended Sediment in Closed Channels," Trans., ASCE, vol. 117, p. 409, 1952.

Juliano, D. W., "Reaeration Measurements in an Estuary," J. San. Engr. Div., ASCE, vol. 95, no. SA6, pp. 1165-1178, Dec. 1969.

Kanwisher, J., "On the Exchange of Gases between the Atmosphere and the Sea," Deep-Sea Research, vol. 10, pp. 195-207, 1963A.

Kanwisher, J., "Effect of Wind on CO_2 Exchange Across the Sea Surface," J. Geoph. Res., vol. 68, no. 13, pp. 3921-3927, July 1, 1963B.

Kehr, R. W., "Effect of Sewage on Atmospheric Reaeration Rates under Stream Flow Conditions," Sewage Works J., vol. 10, no. 2, p. 228, 1938.

Keshavan, K., G. C. Sornberger, and C. I. Hirshberg, "Oxygen Sag Curve with Thermal Overload," J. Envir. Engr. Div., ASCE, vol. 99, no. EE5, pp. 569-575, Oct. 1973.

Kishinevsky, M. Kh., "The Theoretical Work of Danckwerts in the Field of Absorption," J. of Applied Chem. of the USSR (English Translation), vol. 27, no. 4, pp. 359-365, April 1954.

Kishinevsky, M. Kh. and V. T. Serebryansky, "The Mechanism of Mass Transfer at the Gas-Liquid Interface with Vigorous Stirring," J. of Applied Chem. of the USSR (English Translation), vol. 29, no. 1, pp. 29-33, Jan. 1956.

Kramer, G. R., "Predicting Reaeration Coefficients for Polluted Estuary," J. Envir. Engr. Div., ASCE, vol. 100, no. EE1, pp. 77-92, Feb. 1974.
Krenkel, P. A., "Turbulent Diffusion and the Kinetics of Oxygen Absorption," Ph.D Thesis, University of California, Berkeley, 1960.
Krenkel, P. A. and G. T. Orlob, "Turbulent Diffusion and the Reaeration Coefficient," Trans., ASCE, vol. 128, part III, pp. 293-334, 1963.
Lau, Y. L., "Prediction Equation for Reaeration in Open-Channel Flow," J. San. Engr. Div., ASCE, vol. 98, no. SA6, pp. 1063-1968, Dec. 1972A.
Lau, Y. L., "A Review of Conceptual Models and Prediction Equations for Reaeration in Open-Channel Flow," Technical Bulletin 61, Inland Waters Branch, Department of the Environment, Ottawa, Canada, 28 p., 1972B.
Lau, Y. L., "Turbulent Surface Film Thickness for Oxygen Absorption in Open Channel Flows," Hydr. Research Div., Canada Centre for Inland Waters, Burlington, Ont., 16 p., March 1975A.
Lau, Y. L., "An Experimental Investigation of Reaeration in Open Channel Flow," Progress in Water Tech., Pergamon Press, vol. 7, nos. 3/4, pp. 519-530, 1975B.
Lewis, W. K. and W. G. Whitman, "Principles of Gas Absorption," Ind. and Engr. Chem., vol. 16, no. 12, pp. 1215-1220, Dec. 1924.
Lin, S. H., L. T. Fan, and C. L. Hwang, "Design of the Optimal Outfall System for a Stream Receiving Thermal and Organic Waste Discharges," Water Research, vol. 9, no. 7, pp. 623-630, July 1975.
Mattingly, G. E., "Experimental Study of Wind Effects on Reaeration," J. Hydr. Div., ASCE, vol. 103, no. HY3, pp. 311-323, March 1977.
Metzger, I., "Effects of Temperature on Stream Aeration," J. San. Engr. Div., vol. 94, no. SA6, pp. 1153-1159, Dec. 1968.
Micka, T., E. R. Holley, and F. W. Sollo, "Reaeration Experiments with Microorganisms," J. Envir. Engr. Div., ASCE, vol. 99, no. EE6, pp. 971-975, Dec. 1973.
Miyamoto, S. et al, "A Theory of the Rate of Solution of Gas into Liquid," Bulletin, Chem. Soc. of Japan, vol. 5, p. 123ff, 1931.
Morris, J. C. et al, discussion of "Solubility of Atmospheric Oxygen in Water," J. San. Engr. Div., ASCE, vol. 87, no. SA1, pp. 81-86, Jan. 1961.
Negulescu, M. and V. Rojanski, "Recent Research to Determine Reaeration Coefficient," Water Research, vol. 3, no. 3, pp. 189-202, 1969.
O'Connor, D. J., "The Measurement and Calculation of Stream Reaeration Ratio," Oxygen Relationships in Streams, Tech. Report W-58-2, Taft Sanitary Engineering Center, 1958.
O'Connor, D. J. and W. E. Dobbins, "Mechanism of Reaeration in Natural Streams," Trans., ASCE, vol. 123, pp. 641-666, 1958.
Orlob, G. T., "Eddy Diffusion in Homogeneous Turbulence," Trans., ASCE, vol. 126, part I, pp. 397-438, 1959.

Owens, M., R. W. Edwards, and J. W. Gibbs, "Some Reaeration Studies in Streams," Intern. Journ. Air and Water Poll., vol. 8, pp. 469-486, 1964.

Pasveer, A., "Research on Activated Sludge--I. A Study of the Aeration of Water," Sewage and Ind. Wastes, vol. 25, no. 11, pp. 1253-1258, Nov. 1953.

Poon, C.P.C., and H. Campbell, "Diffused Aeration in Polluted Water," Water and Sew. Works, vol. 114, pp. 461-463, 1967.

Rathbun, R. E., discussion of "Wind Shear-Turbulence and Reaeration Coefficient," J. Hydr. Div., ASCE, vol. 98, no. HY9, pp. 1733-1735, Sept. 1972.

Rathbun, R. E., "Reaeration Coefficients of Streams — State-of-the Art," J. Hydr. Div., ASCE, vol. 103, no. HY4, pp. 409-424, April 1977.

Rood, O. E. and E. R. Holley, "Critical Oxygen Deficit for a Bank Outfall," J. Envir. Engr. Div., ASCE, vol. 100, no. EE3, June 1974.

Sawyer, C. N. and P. L. McCarty, Chemistry for Sanitary Engineers, 2nd ed., McGraw-Hill, 518 p., 1967.

Sornberger, G. C. and K. Keshavan, "Simulation of Dissolved Oxygen Profile," J. Envir. Engr. Div., ASCE, vol. 99, no. EE4, pp. 479-488, Aug. 1973.

Streeter, H. W. and E. B. Phelps, "A Study of the Pollution and Natural Purification of the Ohio River--III. Factors Concerned in the Phenomena of Oxidation and Reaeration," U.S. Public Health Serv., Publ. Health Bull. 146, 76 p., 1925.

Thackston, E. L., "Longitudinal Mixing and Reaeration in Natural Streams," Ph.D. Thesis, Vanderbilt Univ., Nashville, 1966.

Thackston, E. L. and P. A. Krenkel, "Reaeration Prediction in Natural Streams," J. San. Engr. Div., ASCE, vol. 95, no. SA1, pp. 65-94, Feb., 1969.

Thames Survey Committee and the Water Pollution Research Laboratory, "Effects of Polluting Discharges on the Thames Estuary," Her Majesty's Stationery Office, London, pp. 357-358, 1964.

Tsao, G. T., "Simultaneous Gas-Liquid Interfacial Oxygen Absorption and Biochemical Oxidation," Biotech. and Bioengr., vol. 10, pp. 1289-1290, 1968.

Tsivoglou, E. C. and J. R. Wallace, "Characterization of Stream Reaeration Capacity," Report EPA-R3-72-012, Project 16050 EDT, U.S. Government Printing Office, Washington, 317 p., Oct. 1972.

van der Kroon, G.T.M., "The Influence of Suspended Solids on the Rate of Oxygen Transfer in Aqueous Solutions," Water Research, vol. 2, pp. 26-30, 1968.

Vanoni, V. A., "Transportation of Suspended Sediment by Water," Trans., ASCE, vol. 111, p. 67, 1946.

Washburn, E. W., International Critical Tables, vol. 5, pp. 63-69 1929.

Whitman, W. G., "The Two-Film Theory of Gas Absorption," Chem. and Met. Engr., vol. 29, no. 4, pp. 146-148, July 23, 1923.

Wilson, G. T. and N. Macleod, "A Critical Appraisal of Empirical Equations and Models for the Prediction of the Coefficient of Reaeration of Deoxygenated Water," Water Research, vol. 8, pp. 341-366, 1974.

8. NOTATION

A = surface area
a = ratio of surface area to volume
C = depth averaged concentration of dissolved oxygen
c = temporal mean concentration
c_s = saturation concentration
D = saturation deficit
D_o = initial dissolved oxygen deficit
D^* = diffusion coefficient
D_t^* = turbulent diffusivity
D_m^* = molecular diffusivity
f = Darcy-Weisbach friction factor
$f(t)$ = proportional part of the vertical elements of water
F_r = Froude number
h = average depth
K_L = liquid film coefficient or transfer coefficient
K_2 = reaeration coefficient = $2.30\ k_2$
p = partial pressure of oxygen
q = rate of mass transport through surface area A
r = average rate at which vertical mixing takes place
R_e = Reynolds number
S_c = Schmidt number
T = temperature
t = time
t' = average time between complete mixings
U = average water velocity
U_a = wind speed
u_{*o} = shear velocity on channel boundary = $\sqrt{\tau_o/\rho}$
u_{*s} = shear velocity on water surface due to wind = $\sqrt{\tau_s/\rho_a}$
V = volume of water
y = vertical coordinate
α = ratio of K_L values
α' = Henry's Law constant
ν = molecular kinematic viscosity
θ = temperature coefficient, Eq. (5)
δ = film thickness
ρ_o = density of water
ρ_a = density of air
τ_o = boundary shear stress
τ_s = shear stress on water surface.

DISSIPATIVE PHYSICO-CHEMICAL TRANSPORT IN THE PYCNOCLINE REGION OF THE OCEAN

John Gribik
Fletcher Osterle

Basic Technology, Inc.
Carnegie-Mellon University

ABSTRACT

This paper presents an analysis of the vertical structure of the ocean and the dissipative processes occurring in it. The analysis is based on thermodynamic principles used in conjunction with a suitable physico-chemical model of sea water. The limitations of this model are thoroughly discussed and particular attention is given to the manner in which the chemical potentials of the sea water constituents must be modified when they are applied to systems of geophysical scale.

The resulting theory is applied to a three-layer model of the nonequilibrium ocean. The model consists of a shallow surface layer which is separated from the underlying deep water by the pycnocline region. The pycnocline acts as an effective barrier to gross vertical mixing and coupled heat and mass diffusion processes operate to modify the solute concentrations. The resulting concentration gradients are compared with those for the limiting case of an equilibrium ocean and the total dissipation is estimated.

Introduction

Sea water is a moderately concentrated solution of many different electrolytic and nonelectrolytic solutes. In his discussion of the thermodynamics of sea-water, Horne (1969) describes sea water as a "horrendous complex soup of just about everything imaginable." Horne goes on to observe that we have essentially no theoretical understanding of mixed electrolyte solutions and that the proposed models of concentrated single electrolyte solutions are little better than curve fits to rather sparse experimental data. On the other hand, he notes that the theoretical models of ideal and near ideal (or dilute) solutions are adequate for many purposes. He suggests that sea water can be represented by a 0.6 molar NaCl solution, and that any theoretical discussion of its thermodynamics should be expressed in terms of a dilute solution model. At present there seems to be no workable alternative. In this paper we will use a consistently applied dilute solution model to examine the vertical structure of equilibrium and nonequilibrium oceans.

The Equilibrium Criteria

A multi-component homogeneous system is in thermodynamic equilibrium when the temperature and the individual component chemical potentials are uniform throughout. If we assume for the present analysis that the ocean is essentially one-dimensional and define the depth coordinate, y, to increase downward from the surface, the equilibrium conditions are:

$$\frac{dT}{dy} = 0 \tag{1}$$

$$\frac{d\tilde{\mu}_i}{dy} = 0; \qquad i = 0,1,\ldots n \tag{2}$$

where T is the absolute temperature and $\tilde{\mu}_i$ is the electrochemical potential related to the chemical potential (or partial molar free enthalpy), μ_i, by

$$\tilde{\mu}_i = \mu_i + z_i F \phi \tag{3}$$

where z_i is the charge number (positive for cations, negative for anions, and zero for a neutral component), F is Faraday's constant, and ϕ is the electrical potential.

The scale on which the ocean is treated is geophysical, so that gravity cannot be neglected. Gravitational potential energy ($-M_i gy$) is included in the chemical potential, where M_i is the molecular mass of the component and g the gravitational constant.

The Dilute Solution Model

In a dilute solution, solvent molecules far outnumber solute molecules, which suggests that solvent-solute interactions will overwhelm solute-solute interactions. The dilute solution model views the solute as a collection of discrete particles suspended in a continuum (the solvent). The particles can be ions or neutral molecules. In our ocean model we will be concerned with several chemical species. They are identified by the following subscripts:

$i = 0$; water (the solvent)
$i = 1$; Na^+ (cation)
$i = 2$; Cl^- (anion)
$i = 3$; O_2 (dissolved)
$i = 4$; N_2 (dissolved)

In dilute solution theory the solution pressure is the sum of the partial pressures of the components, represented by

$$p = p_o + \sum_{i=1}^{n} p_i \tag{4}$$

where the solute pressures are given by the van't Hoff relation

$$p_i = c_i RT \tag{5}$$

with c_i the solute concentrations (moles per unit volume of solution) and R the gas constant. The molar volume, v, of the solution is given by

$$v = \bar{v}_o \left(1 - \sum_{i=1}^{n} x_i\right) + \sum_{i=1}^{n} \bar{v}_i x_i \tag{6}$$

where the x_i's are the solute mole fractions, and the $\bar{v}_i$'s are the partial molar volumes of the solution components. The partial molar volumes of relatively small molecules (e.g., water, salts, gases) are of the same order of magnitude, so that for dilute solutions, equation (6) can be approximated by

$$v = \bar{v}_o \tag{7}$$

Now the chemical potential of the solvent is given by

$$\mu_o = g_o - vRT \sum_{i=1}^{n} c_i = M_o gy \tag{8}$$

where g_0 is the molar free enthalpy of pure water at the solution pressure and temperature. The second term corrects for the fact that the solvent "sees" a pressure equal to the solution pressure less the sum of the solute pressures. The electrochemical potentials of the solutes are given by

$$\tilde{\mu}_i = \mu_i^* + RT\ell nc_i + z_i F\phi - M_i gy \qquad (9)$$

where μ_i^* is the chemical potential of the solute in a hypothetical ideal solution of unit molarity at the solution pressure and temperature. The second term corrects for the fact that the solute concentration is not normal.

Equilibrium Pressure and Salt Concentration Gradients

The concentration of salt in the ocean is approximately 1000 times greater than that of the principal dissolved gases. The gases will be neglected for the moment since their effect on the chemical potential of the ocean must be minor. The differential equations governing the pressure and salt concentration gradients are obtained by combining the equilibrium criteria (equation (2)) with the chemical potential relations (equations (8), (9)). They are

$$vdp - 2vRTdc_s - M_0 gdy = 0 \qquad (10)$$

$$\bar{v}_i dp + \frac{RT}{c_s} dc_s + Fd\phi - M_i gdy = 0 \qquad (11)$$

$$\bar{v}_2 dp + \frac{RT}{c_s} dc_s - Fd\phi - M_2 gdy = 0 \qquad (12)$$

for the solvent (water), cation (Na^+), and anion (Cl^-) respectively. The first terms are determined by the fundamental relationship

$$\bar{v}_i = \left(\frac{\partial \mu_i}{\partial p}\right)_{c_i, T} \qquad (13)$$

and the electroneutrality condition, $c_1 = c_2 = c_s$, has been invoked with c_s the salt concentration. The electrical potential gradient can be eliminated by adding equations (11) and (12) to obtain

$$\bar{v}_s dp + \frac{2RT}{c_s} dc_s - M_s gdy = 0 \qquad (14)$$

where

$\bar{v}_s = \bar{v}_1 + \bar{v}_2$, the salt partial molar volume.

$M_s = M_1 + M_2$, the salt molecular mass.

Now solving equations (10) and (14) for the pressure gradient, we have

$$dp = \frac{g}{v}\left[\frac{M_o + c_s v M_s}{1 + c_s \bar{v}_s}\right] dy \qquad (15)$$

It will be useful to compare equation (15) with the pressure gradient expression which results from hydrostatics. Assuming that the ocean is in hydrostatic equilibrium, we have

$$dp = \frac{g}{v}\left[M_o + c_s v(M_s - M_o)\right] dy \qquad (16)$$

This expression for the pressure gradient is exact. The apparent discrepancy between equations (15) and (16) is resolved when we expand the denominator in equation (15) as

$$\frac{1}{1 + c_s \bar{v}_s} = 1 - c_s \bar{v}_s + (c_s \bar{v}_s)^2 + \ldots \qquad (17)$$

and retain only the first two terms of the series, as a measure of the degree of dilution for which we expect dilute solution theory to be valid. Equation (15) then becomes

$$dp = \frac{g}{v}\left[M_o + c_s(vM_s - \bar{v}_s M_o)\right] dy \qquad (18)$$

on the condition that $(c_s \bar{v}_s)^2 << 1$. Equation (18) is correct to the extent that $\bar{v}_s$ equals v. The error is small since for NaCl, $\bar{v}_s$ is only 5% greater than v. We will use equation (18) in preference to the exact expression since we wish to explore the implications of applying a consistent dilute solution model to the ocean.

Next we take up the question of the salt concentration gradient. Eliminating the pressure between equations (10) and (18), we obtain

$$dc_s = \frac{c_s g}{2RT} (M_s - \frac{\bar{v}_s}{v} M_o) \tag{19}$$

from which

$$c_s = c_{so} \exp[\frac{g(M_s - \frac{\bar{v}_s}{v} M_o)y}{2RT}] \tag{20}$$

where c_{so} is the salt concentration at the surface. At a depth of 1000 meters we find that $c_s = 1.08\ c_{so}$ indicating that the effect of gravity on the salt concentrations is relatively small to depths of this magnitude.

The Dissolved Gases

To conclude our discussion of the equilibrium ocean we will now consider the ocean-atmosphere equilibrium state. The governing differential equation for the dissolved gas concentrations in the ocean is obtained by combining the equilibrium criteria with the expressions for the chemical potential of neutral solutes. There results

$$dc_i = \frac{c_i}{RT} [M_i g - \bar{v}_i \frac{dp}{dy}] dy , \quad i = 3, 4 \tag{21}$$

Substituting for the pressure gradient from equation (18) and for the concentration from equation (20) written for the gases, and integrating, we obtain

$$c_i = c_{io} \exp[\frac{g}{RT} [M_i - \frac{\bar{v}_i}{v} M_o] y - \frac{2RT\bar{v}_i}{g} (c_s - c_{so})] , \quad i = 3, 4 \tag{22}$$

At 1000 meters the ratio of oxygen and nitrogen concentrations to their values at the surface are 0.997 and 0.978 respectively. The equilibrium concentrations decrease slightly with depth.

The Deep Water Dissipation

The vertical structure of the open ocean can be divided into three distinct zones. These are the surface, pycnocline, and deep zones (Gross, 1971). The surface zone is a relatively shallow, well mixed layer which is strongly affected by insolation.

Absorbed solar radiation warms the surface water, causing it to be less dense than the underlying water. Depending on the relative amounts of evaporation and precipitation, the surface water may be more or less saline than the deep water. Because the surface water is well mixed, its density is essentially uniform and it exists in near neutral stability; i.e., water particles may move rather easily in the vertical direction. Below the surface zone lies the pycnocline in which the water density increases with depth. This density gradient is caused by a temperature gradient (the thermocline) and a salinity gradient (the halocline) acting in conjunction. Because of the density gradient, the water in the pycnocline has great stability, and the pycnocline acts as an effective barrier to vertical water movement thus providing a floor for surface circulation and seasonal changes in temperature and salinity.

As the sun-warmed ocean surface water flows slowly from the equator to the poles, it is intensely cooled and its density consequently increased. The surface water gradually sinks at the high latitudes to become the water of the deep zone which flows slowly back to the equator. Over most of the ocean, the deep water is isolated from the surface water by the pycnocline, and the temperature of the deep water remains low (averaging about 3.5 deg C) as well as the salinity. Since there are temperature and salinity gradients between the surface water and the deep water, there will be dissipatory processes (conduction and diffusion) occurring in the pycnocline zone. We wish to examine the nature of these processes and determine the magnitude of the dissipation occurring in the pycnocline. The analysis will be performed within the framework of the dilute solution model discussed above and, in particular, will be subject to the following assumptions:

1) the ocean is in local (but not global) thermodynamic equilibrium
2) the pycnocline is in mechanical equilibrium
3) all transport processes in the pycnocline are one dimensional and molecular in nature
4) the pycnocline is an isotropic region in steady state
5) the dissipation caused by the diffusion of dissolved gases can be neglected.

The Local Dissipation Equation

The proper forces upon which the phenomenological conduction and diffusion equations must be based are determined by the form of the local dissipation equation. The dissipation equation follows from the application of the first and second laws of thermodynamics to an elemental control volume in the medium.

The first law reads:

$$- \frac{d\dot{Q}}{dy} = \frac{d}{dy} \sum_{i=0}^{2} J_i(\bar{h}_i + z_i f\phi - M_i gy) \tag{23}$$

where $\dot{Q}$ is the conduction heat flux, J_i and $\bar{h}_i$ are, respectively, the molar flux and partial molar enthalpy of the i'th constituent.

The second law reads:

$$- \frac{d}{dy} (\frac{\dot{Q}}{T}) + \dot{\theta} = \frac{d}{dy} \sum_{i=0}^{2} J_i \bar{s}_i \tag{24}$$

where $\dot{\theta}$ is the volumetric entropy production rate, and $\bar{s}_i$ is the partial molar entropy of the i'th constituent.

Multiplying equation (24) by a reference temperature, T_o, and subtracting the resulting equation from equation (23), we obtain

$$T_o \dot{\theta} = (\frac{T_o}{T} - 1) \frac{d\dot{Q}}{dy} - (\frac{T_o}{T^2} \frac{dT}{dy}) \dot{Q} + \frac{d}{dy} \sum_{i=0}^{2} J_i(- \tilde{\mu}_{oi}) \tag{25}$$

where $T_o \dot{\theta}$ is the total dissipation, and $\tilde{\mu}_{oi}$ is a modified electrochemical potential defined by

$$\tilde{\mu}_{oi} = \bar{h}_i - T_o \bar{s}_i + z_i F\phi - M_i gy \tag{26}$$

and related to the conventional electrochemical potential by

$$\tilde{\mu}_{oi} = \tilde{\mu}_i - (T_o - T) \bar{s}_i \tag{27}$$

The last term in equation (25) can be expanded as follows:

$$\sum_{i=0}^{2} J_i(- \frac{d\tilde{\mu}_{oi}}{dy}) + \sum_{j=1}^{r} A_{oj} \Omega_j \tag{28}$$

where A_{oj} and Ω_j are the modified chemical affinity and the reaction rate of the j'th chemical reaction.

The equation for the local dissipation follows from equation (25) by letting T_o go to T, the local absolute temperature. From the thermodynamic relation

$$\left(\frac{\partial \tilde{\mu}_i}{\partial T}\right)_{p,c_1} = -\bar{s}_i$$

we see that

$$\frac{d\tilde{\mu}_{oi}}{dy} = \left(\frac{d\tilde{\mu}_i}{dy}\right)_T$$

where, on the right, the bracket and the subscript T denotes the derivative taken at constant temperature. Also as T_o goes to T, A_{oj} and $\tilde{\mu}_{oi}$ go to their conventional values A_j and $\tilde{\mu}_i$. If we now recognize that in the pycnocline the bulk or mass average is zero, the local dissipation finally takes the form

$$T\dot{\theta} = \dot{Q}\left(-\frac{1}{T}\frac{dT}{dy}\right) + \sum_{i=1}^{2} J_i\left(-\frac{d\tilde{\mu}_i}{dy}\right)_T + \sum_{j=1}^{r} \Omega_j A_j \quad (29)$$

The fluxes and forces characterizing the dissipative processes occurring in the pycnocline region are identified by equation (29). In the bilinear form of equation (29) the first terms are fluxes and the second terms forces. It is reasonable to expect that the fluxes can be related to the forces by a set of linear phenomenological equations. The scalar chemical affinities will, of course, not appear in the vector heat flux and molar diffusion equations since the Curie symmetry principle (Haase, 1969) prohibits such coupling in isotropic media. In any event, the contribution to the dissipation in the pycnocline from whatever chemical reactions which might be occurring there will be neglected.

The Phenomenological Equations

The anticipated coupling equations can be written as:

$$J_1 = L_{11}\left(-\frac{d\tilde{\mu}_1}{dy}\right) + L_{12}\left(-\frac{d\tilde{\mu}_2}{dy}\right) + L_{13}\left(-\frac{1}{T}\frac{dT}{dy}\right) \quad (30)$$

$$J_2 = L_{21}\left(-\frac{d\tilde{\mu}_1}{dy}\right) + L_{22}\left(-\frac{d\tilde{\mu}_2}{dy}\right) + L_{23}\left(-\frac{1}{T}\frac{dT}{dy}\right) \quad (31)$$

$$\dot{Q} = L_{31}\left(-\frac{d\tilde{\mu}_1}{dy}\right) + L_{32}\left(-\frac{d\tilde{\mu}_2}{dy}\right) + L_{33}\left(-\frac{1}{T}\frac{dT}{dy}\right) \quad (32)$$

where the L's are the coupling coefficients. The Onsager reciprocal theorem would lead us to expect that near equilibrium, $L_{ij} = L_{ji}$.

We will now evaluate these coefficients from a simple dilute solution transport model using when necessary the Onsager reciprocal theorem.

According to this model the ions in a dilute solution are, in general, subject to the following forces (on a mole basis).

1) an osmotic force resulting from the gradient in ion partial pressure

$$-\frac{RT}{c_i}\frac{dc_i}{dy}$$

2) an electrical body force

$$-z_i F\frac{d\phi}{dy}$$

3) a gravitational force

$$M_i g$$

4) a bouyancy force

$$-\bar{v}_i\rho_o g$$

where ρ_o is the solvent density

5) a drag force exerted by the solvent

$$-\frac{u_i - u_o}{m_i}$$

where m_i is the mobility per mole of ions.

Neglecting inertia forces, a force balance leads to

$$-\bar{v}_i\rho_o g - \frac{RT}{c_i}\frac{dc_i}{dy} - z_i F\frac{d\phi}{dy} + gM_i - \frac{u_i - u_o}{m_i} = 0 \qquad (33)$$

Neglecting u_o, and solving for the molar ionic flux, we obtain:

$$J_i = c_i u_i = m_i c_i[-\bar{v}_i\rho_o g - \frac{RT}{c_i}\frac{dc_i}{dy} - z_i F\frac{d\phi}{dy} + M_i g] \qquad (34)$$

If the solution density can be approximated by the solvent density, we have

$$\frac{dp}{dy} = \rho_o g$$

and

$$J_1 = m_i c_i[-\bar{v}_i \frac{dp}{dy} - \frac{RT}{c_i}\frac{dc_i}{dy} - z_i F \frac{d\phi}{dy} + M_i g] = m_i c_i(- \frac{d\tilde{\mu}_i}{dy})_T \quad (35)$$

We have shown therefore that $L_{11} = m_1 c_s$, $L_{12} = L_{21} = 0$, $L_{22} = m_2 c_s$. The coefficient L_{i3} $(i = 1,2)$ is empirically determined and usually expressed in terms of the Soret coefficient, σ_i, according to

$$L_{i3}(- \frac{1}{T}\frac{dT}{dy}) = D_i \sigma_i c_i \frac{dT}{dy} \quad (36)$$

where the diffusivity, D_i, is related to the mobility, m_i, by Einstein's relation

$$D_i = m_i RT \quad (37)$$

The coefficient L_{33} is clearly related to the ordinary coefficient of thermal conductivity, λ, by $L_{33} = \lambda T$. Finally the coefficients L_{31} and L_{32} are determined by Onsager's reciprocal theorem.

Equations (30), (31), (33) with their coupling coefficients now known can be rearranged to a more useful form, namely:

$$J_i = -D_i c_s[\frac{1}{c_s}\frac{dc_s}{dy} + \frac{F}{RT}\frac{d\phi}{dy} - \frac{g}{RT}(M_i - M_o \frac{\bar{v}_i}{v}) - \sigma_i \frac{dT}{dy}] , \quad i = 1,2 \quad (38)$$

$$\dot{Q} = (D_1\sigma_1 + D_2\sigma_2) RT^2 \frac{dc_s}{dy} + (D_1\sigma_1 - D_2\sigma_2) c_s TF \frac{d\phi}{dy} - D_1\sigma_1 c_s Tg(M_1 - M_o \frac{\bar{v}_1}{v}) - D_2\sigma_2 c_s Tg(M_2 - M_o \frac{\bar{v}_2}{v}) - \lambda \frac{dT}{dy} \quad (39)$$

If there is no electrical current in the vertical direction, the electrical potential (diffusion potential) gradient will be such that $J_1 = J_2$. The salt flux defined by

$$J_s = \frac{1}{2}(J_1 + J_2)$$

and the heat flux now becomes

$$J_s = \frac{-2D_1D_2}{D_1 + D_2}\frac{dc_s}{dy} + \frac{D_1D_2}{D_1 + D_2}\frac{gc_s}{RT}(M_s - M_o\frac{\bar{v}_s}{v}) + \frac{D_1D_2(\sigma_1 + \sigma_2)}{D_1 + D_2}c_s\frac{dT}{dy} \tag{40}$$

$$\dot{Q} = [\frac{(D_1\sigma_1 - D_2\sigma_2)^2}{D_1 + D_2}c_sRT^2 - \lambda]\frac{dT}{dy} + \frac{2D_1D_2(\sigma_1 + \sigma_2)RT^2}{D_1D_2}\frac{dc_s}{dy}$$
$$- \frac{D_1D_2(\sigma_1 + \sigma_2)}{D_1 + D_2}c_sTg(M_s - M_o\frac{\bar{v}_s}{v}) \tag{41}$$

We see from equation (40) that salt flow is driven by a temperature gradient in addition to a concentration gradient and buoyancy. To see how large a driving force the temperature gradient can be we assumed a linear temperature profile from 298°K at a depth of 1000 meters and calculated the concentration difference which would just balance this driving force in the sense of reducing the salt flow to zero. The result is that a 20°C linear decrease in temperature has the same effect on salt diffusion as a 3% increase in salt concentration. Recalling that in an equilibrium sea the salt concentration at a depth of 1000 meters is 8% above the value at the surface, we see that the thermocline will augment the flow of salt from the surface down the halocline in those areas of the ocean where the salinity at depth is less than at the surface.

The Pycnocline Dissipation

To determine the dissipation in the pycnocline, we consider a control volume which encloses the entire zone. Assuming steady state and no chemical reactions, the total or global dissipation per unit of ocean surface area was determined for the following data:

surface temperature = 298°K
surface salt concentration = 0.63 moles/liter
temperature at 1000m = 278°K
salt concentration at 1000m = 0.61 moles/liter

The thermal dissipation was found to completely dominate the contributions from salt diffusion. The magnitude of the dissipation worked out to be about 8×10^{-4} watts/m^2, which is extremely small when compared to the incoming solar energy flux which averages approximately 150 watts/m^2 at the ocean surface (Brahtz, 1968). However, the total pycnocline dissipation is still quite large. The area of the earth's surface is 510 x 10^6 square kilometers of which 361 x 10^6 square kilometers is the surface area of the oceans and seas. If we assume that a pycnocline zone exists over one-half of this area and that the temperature and concentration gradients used here are representative, the total dissipation would be about 0.3 x 10^{12} watts.

To put this number in perspective, we can compare it with the estimated dissipation rates of some other geophysical processes. Surface winds transfer kinetic energy to the ocean at the rate of about 10^{14} watts. Part of this energy is stored in surface waves and is dissipated along the ocean shorelines. Inman and Brush (1973) estimate that this mechanism dissipates energy at the rate of about 2.5 x 10^{12} watts. Strong tidal currents are also a significant source of dissipation in the ocean. The principal tidal dissipation occurs in the shallow areas of the ocean, and the total rate of energy dissipation by lunar and solar tides has been estimated at 2.2 x 10^{12} watts (Inman and Brush, 1973). Finally, the dissipation associated with the mixing of all the relatively salt-free river water draining into the sea with the saline sea has been estimated at about 2 x 10^{12} watts (Weinstein and Leitz, 1976).

References

Horne, R. A., Marine Chemistry, New York; Wiley-Interscience, 1969.

Gross, M. G., Oceanography, 2nd ed., Columbus; Charles E. Merrill, 1971.

Haase, R., Thermodynamics of Irreversible Processes, trans., Menlo Park; Addison-Wesley, 1969.

Brahtz, J. F., Ocean Engineering, New York; John Wiley & Sons, Inc. 1968.

Inman, D. L. and B. M. Brush, "The Coastal Challenge," Science, 181 (1973), 20-32.

Weinstein, J. N. and F. B. Leitz, "Electric Power from Differences in Salinity: The Dialytic Battery," Science, 191 (1976), 557-559.

A HYDRODYNAMIC MODEL FOR THE TRANSPORT OF A CONSERVATIVE POLLUTANT

T.S. Murty

Marine Environmental Data Service, Ocean and Aquatic Sciences

ABSTRACT

A hydrodynamic model was developed for calculating the transport of a conservative contaminent, and the model has been used for determining the movement of oil slicks in several water bodies in Canada. At present, this model does not consider the structural properties of crude oil (mainly because observations showed that the hydrodynamic forces are at least an order of magnitude greater than the internal spreading forces), but is limited to the dynamics of surface waters. The following forces are included: (1) the effects of periodic or quasi-stationary water currents such as tides or hydraulic channel flows, (2) the effects of organized atmospheric wind systems and the resulting systematic displacements of surface waters, (3) the effects of quasi-random disturbances in wind fields or current patterns with the resulting non-uniform spreading of the oil patch. The topography of the water body, the tidal currents and the climatological data can be stored in a computer data bank, whereas the meteorological data has to be obtained in real time. Secondary factors such as internal spreading forces, stratification and the influence of an ice layer are being included in an approximate manner. In this paper, the discussion is centered on the inclusion of these secondary factors.

INTRODUCTION

Observations on actual oil spills indicate that the primary forces responsible for the movement of oil slicks in a water body are the hydrodynamic forces such as the tidal currents and wind-generated circulation. There are secondary mechanisms such as internal spreading forces (dependent upon viscosity, surface tension and pour point), an ice layer on the water and stratification of the water, which also affect the movement of the slicks.

A completely automated computer model for the prediction of oil slick movement was described by Simons et al. (1975). This model for simulating the movement and dispersion of oil in water did not consider the structural properties of oil but was limited to the dynamics of surface waters in relation to oil slicks. The following factors alone were considered:

1. the effects of periodic or quasi-stationary water currents such as oceanic tides or hydraulic channel flows
2. the effects of organized atmospheric wind systems and the resulting systematic displacements of surface waters
3. the effects of quasi-random disturbances in wind fields or current patterns with the resulting non-uniform spreading of the oil slick.

The above effects have to be dealt with separately for practical reasons. The tidal or hydraulic currents are most readily obtained from observations and are usually stored in oceanographic data centers. The wind-driven flow must be determined from meteorological observations or predictions and require deterministic computer models for water circulation. The random disturbances and dispersion problems are dealt with by statistical methods such as Monte-Carlo simulation.

Thus, Phase One of the computer model ideally should have a data bank for the topography of the water bodies, a second data bank for tidal currents and a third data bank for climatological data. The data needed in real time after the occurrence of a spill are the meteorological conditions.

Figure 1 shows the detailed plan of the Phase One of the computer model. Once the geographic coordinates of the spill are specified, the program selects the appropriate grid (with the relevant hydrographic data). Originally, a great number of particles N, is located at the spill site in the model. The main program now calls for the three subroutines (one for tidal drift, the second for wind drift and the third for dispersion) to determine the displacement of these particles for a time step Δt (taken to be about one hour). The calculation is repeated till the desired number of time steps are completed.

Initially, the influence of tides and wind drift is the same for all particles since they are all at the same starting point. However, the random effects will spread the particles apart and hence, particles at different locations will be effected differently

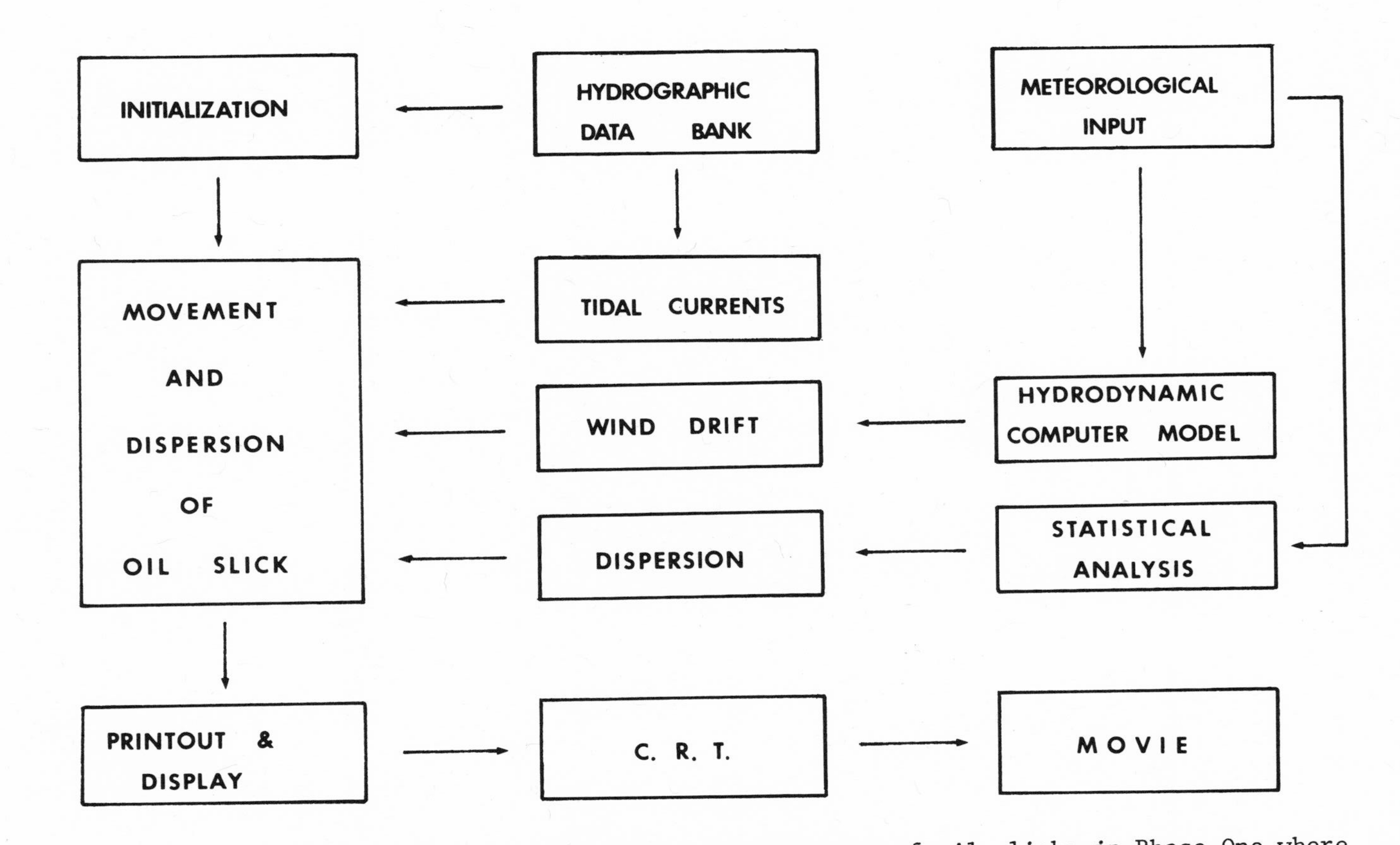

Fig. 1. Simulation plan for calculating the movement of oil slicks in Phase One where only the hydrodynamics is considered.

by tides and wind driven currents. Also, the coastline acts as a barrier to the movement of the particles, perpendicular to the shore. The density of the particle distribution may be taken as the concentration of the oil film on the water surface.

The sections below will discuss the incorporation of secondary factors such as internal spreading force, influence of ice and stratification on the movement of oil slicks. This forms Phase Two of the oil slick simulation model.

INTERNAL SPREADING FORCES

Blokker (1964) developed an empirical relation for the internal or self-spreading (i.e. not considering the external forces) of oil.

$$K_r \ t = \frac{\tau \ (r_t^3 - r_0^3)\rho_w}{3V(\rho_w - \rho_0)\rho_0} \tag{1}$$

where K_r is the Blokker constant (its value depends upon the particular type of oil under consideration), r_0 and r_t are respectively the radii (cm) of the oil slick (assumed to be circular) at times zero at t (expressed in sec), V is the volume of oil spilled (cm^3) and ρ_w and ρ_0 respectively (in units of gm cm^{-3}) are the densities of sea water and oil. This model assumes that the oil spreads uniformly and the rate of spreading decreases exponentially with the reduction in the thickness of the slick.

Fay (1969) used dimensional analysis to deduce the various regimes in the internal spreading of oil slicks. He considered the following forces: gravity, surface tension, inertia, viscosity; out of these, the first two are accelerating forces and the other two are retarding forces. Fay identified the following three regimes and in each regime the radius r of the oil slick at time t is given by the following expressions:

Gravity-inertia regime

$$r = K_1 \ (\Delta\rho \ g \ V \ t^2)^{\frac{1}{4}} \tag{2}$$

Gravity viscous regime

$$r = K_2 \left(\frac{\Delta\rho \ g \ V \ t^2 \ t^{3/2}}{\nu_w^{\frac{1}{2}}} \right)^{1/6} \tag{3}$$

Surface tension-viscous regime

$$r = K_3 \left(\frac{\sigma^2 \ t^3}{\rho_w^2 \ \nu_w} \right)^{\frac{1}{4}} \qquad (4)$$

Here K_1, K_2, K_3 are dimensionless constants whose values respectively are 1.14, 1.45 and 2.30; g is gravity, V is volume of oil (in ft^3) t is time, $\Delta\rho$ is difference of density between water and oil, defined as:

$$\Delta\rho = \frac{\rho_w - \rho_0}{\rho_w} \qquad (5)$$

The surface tension of the oil is σ and its kinematic viscosity of water is ν_w.

Fay has given expressions for the final radius of the slick (i.e. internal spreading has stopped) and also the times at which the various regimes come into effect. For most of the spills, the third regime may be sufficient to describe the spreading due to internal forces. The following relation is generally used to determine the area A of an oil slick (expressed in square nautical miles):

$$A = 2.5\text{x}10^{-5} \ V^{2/3} \ t^{\frac{1}{2}} \qquad (6)$$

Here V is the volume of the oil spilled (gallons) and t is the time (hours) after the initial spill. For a 25,000 ton spill, after one day, equation (6) gives for A, four square nautical miles. In principle, one might use (6) to provide an initial area of the oil slick which then moves and spreads due to the hydrodynamic forces.

One complicating factor could be the dual role played by viscosity (i.e. positive as well as negative). Starr (1968) discussed generally, the negative viscosity concept and states that whereas the effect of positive viscosity is to oppose relative motions in the fluid, the effect of negative viscosity is to sustain differences of velocity or increase these differences. One example Starr gave for negative viscosity is due to the reversal of the shear in the mean flow by tidal currents while the momentum flux remains downward.

Katz et al. (1965) performed field experiments on the response of dye tracers to the surface conditions on the water and found that the dye patches developed curved tails and sometimes they shrank in size. From Ekman theory considerations, the rate of elongation should depend upon the wind speed and the efficiency of the vertical exchange of momentum. Thus, the eddy diffusion, instead of spreading the slick, could shrink it.

Additional observational evidence for the negative viscosity is provided by Assab et al. (1971). They designed a field experiment to explore the effects of different diffusion scales on the spreading of dye patches. Again, the whole dye field shrank which indicates that the horizontal eddy diffusion of the field enclosed by the dye patches was actually negative in the observed scales.

In view of the fact that the viscosity could be positive or negative, it can either play an accelerating role or a retarding role. Then the following question naturally arises: how the three regimes deduced by Fay will be modified. Another point worthy of attention is that some of the crude oils might not behave in a Newtonian sense regarding their viscous behavior and one might be forced to use rheological considerations.

INFLUENCE OF ICE

The influence of ice on the movement of oil slicks can be considered from two different scales: first, the macro-scale, in which we consider the influence of the ice on the slick through the hydrodynamic forces and second, the small-scale in which, how the small slicks may be contained (i.e. pushed under ice patches) when ice and open water co-exist in a water.

First, we consider the macro-scale. Figures 3 and 4 respectively show the movement of the slicks in the summer and winter periods in the Gulf of St. Lawrence (Figure 2) from a spill originating on the Gaspé coast. In the summer period the oil slick moves towards the Prince Edward Island, whereas in the winter period, it moves towards the Anticosti Island. This difference in the trajectory is due to the difference in the weather and wind patterns as well as the surface circulation patterns in the Gulf between the summer and winter periods.

Next, we will consider the influence of ice on oil slicks in the smaller space scales. Observational studies showed that an ice layer on the water can act as an effective barrier to the spreading and movement of an oil slick (e.g. Barber, 1970). Traditionally one uses the concept of Froude number in explaining the containment of oil slicks by booms (e.g. Wilkinson, 1971). Probably the same line of arguments could be extended for containment by ice. However, a different approach is presented below.

There is a difference in the upper boundary condition between

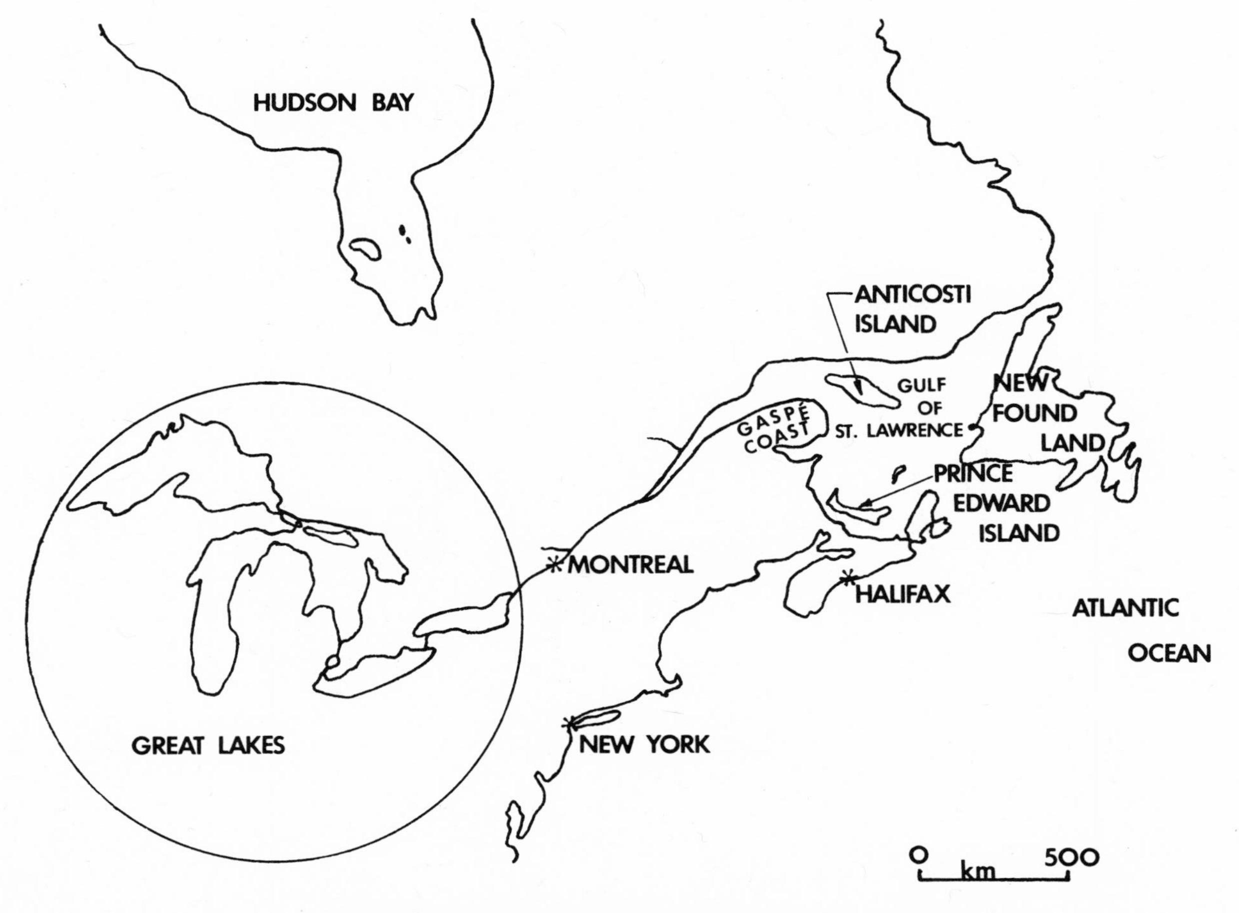

Fig. 2. Geography of the Gulf of St. Lawrence.

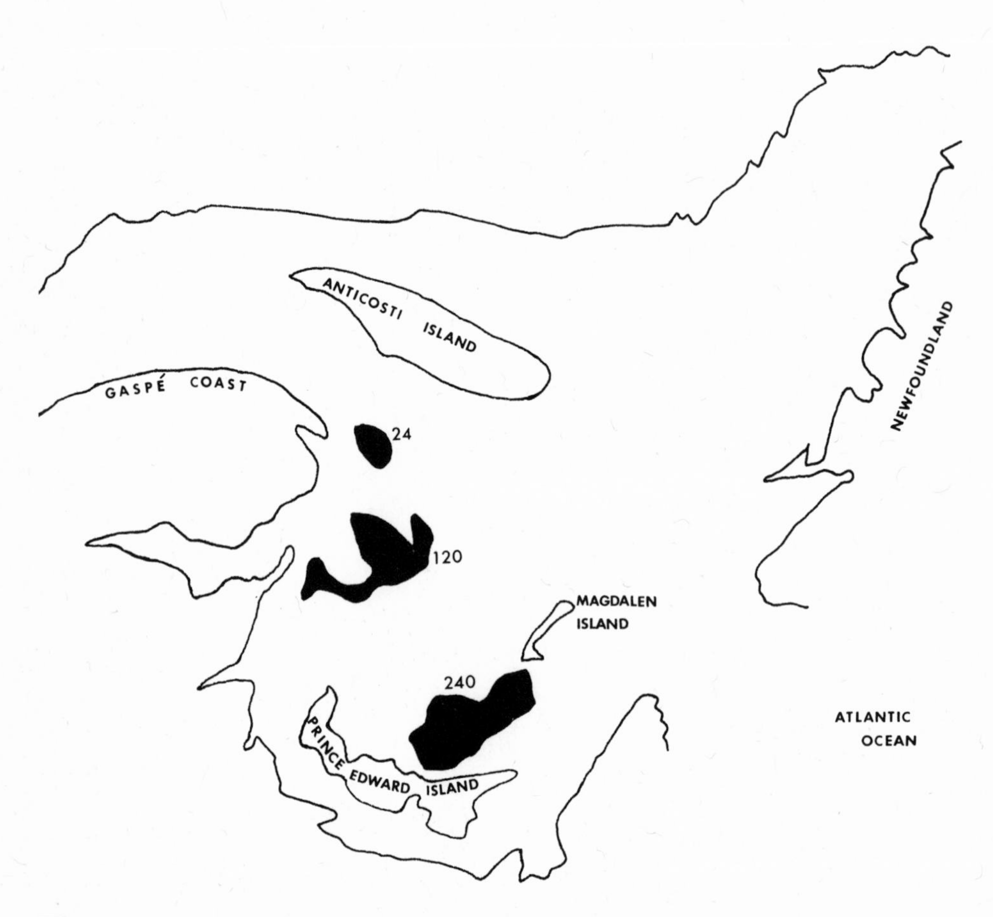

Fig. 3. Oil slick positions at 24, 120 and 240 hours after the initial time of spill for the summer period in the Gulf of St. Lawrence.

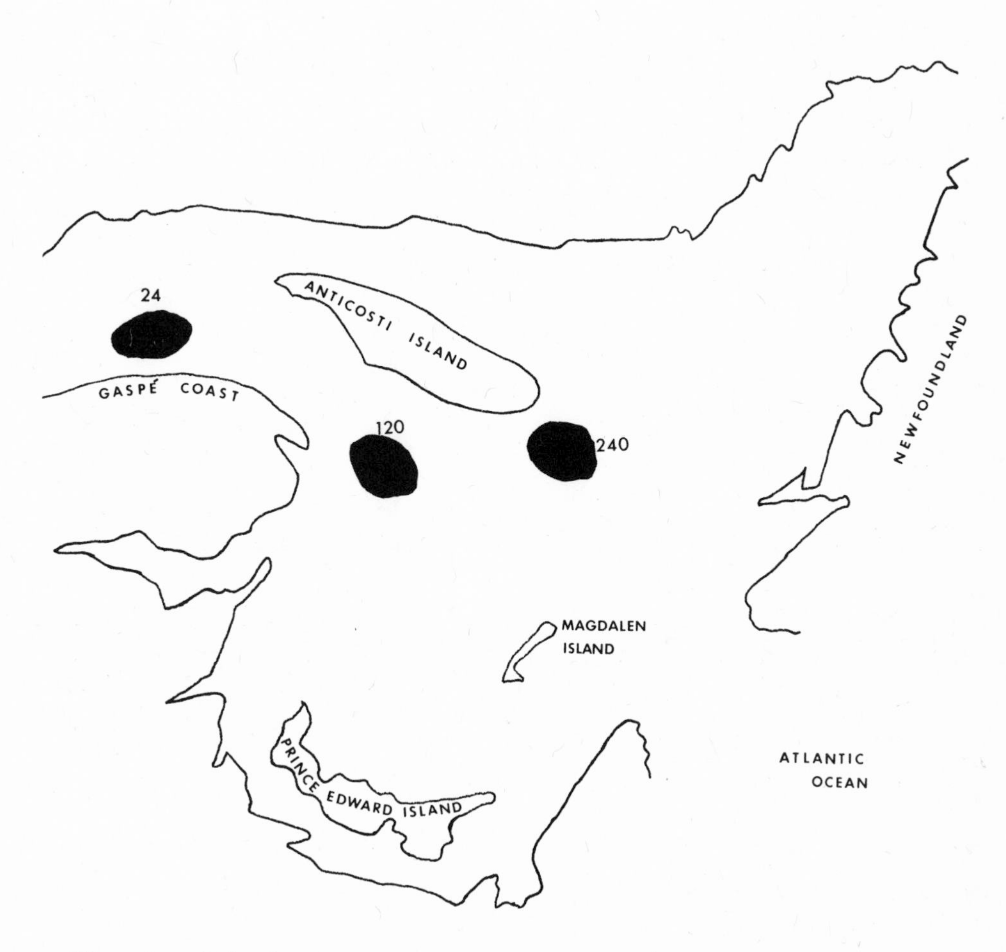

Fig. 4. Oil slick positions at 24, 120 and 240 hours after the initial time of spill for the winter period. In this case, the spill site is somewhat north compared to the spill site for the simulation shown in Figure 1.

ice-covered and ice-free regions, owing to considerations of turbulence (Stewart, 1963). In the ice-free region, the vertical component of motion must (nearly) vanish (there is a higher order non-zero component) and the two horizontal components are unrestricted. In the ice-covered region, all the three components of motion must vanish at the surface.

Thus, the turbulence under an ice cover is isotropic whereas it is anisotropic in the ice-free region. Townsend (1956) showed that in an anisotropic turbulence, there is a net pressure-like stress $(\overline{u^2} - \overline{w^2})$ where $\overline{u^2}$ and $\overline{w^2}$ are the mean square horizontal and vertical turbulent velocity components. Thus, there is a net stress which tends to push the surface water from the ice-free portion, under the ice cover. Longuet-Higgins and Stewart (1962) showed that if the ice-free area is large in extent and if significant wave motion occurs there, then the radiation pressure due to the waves will cause a similar phenomena, i.e. to accumulate surface water under the ice cover. Thus, oil floating on water can be pushed under the ice cover and be contained.

INFLUENCE OF STRATIFICATION

Large oil slicks can occur due to spillage from sunken or grounded tankers following some accident. In this situation, a large quantity of oil may be released at some depth in sea water. Since the density of oil is less than that of sea water, due to buoyancy effects, the oil will tend to rise towards the surface and in this process it could give rise to an important phenomenon known as bifurcated plume.

Hayashi (1971) first examined this phenomenon through laboratory experiments in connection with studies on sewage disposal through pipes placed at the bottom of the sea. The sewage which first forms a bent-over plume, sometimes splits sideways into two concentrated regions with a spacing between them. This means the coastline may be contaminated because about half the effluent (sewage) moves towards the coast, even though the current may be parallel to the coast. Similar danger of contamination exists from oil spilled in water at some depth.

Next, we will look into some details of this bifurcation phenomenon, following Hayashi. With reference to Figure 5, let the oil be discharged as a round buoyant jet into a uniform cross-stream of velocity U. Because the outer part of the plume tends to mix with the water faster than the inner part, a short distance from the outlet, the flow becomes fully developed into a horseshoe shaped profile. The plume becomes bent-over, because the outer

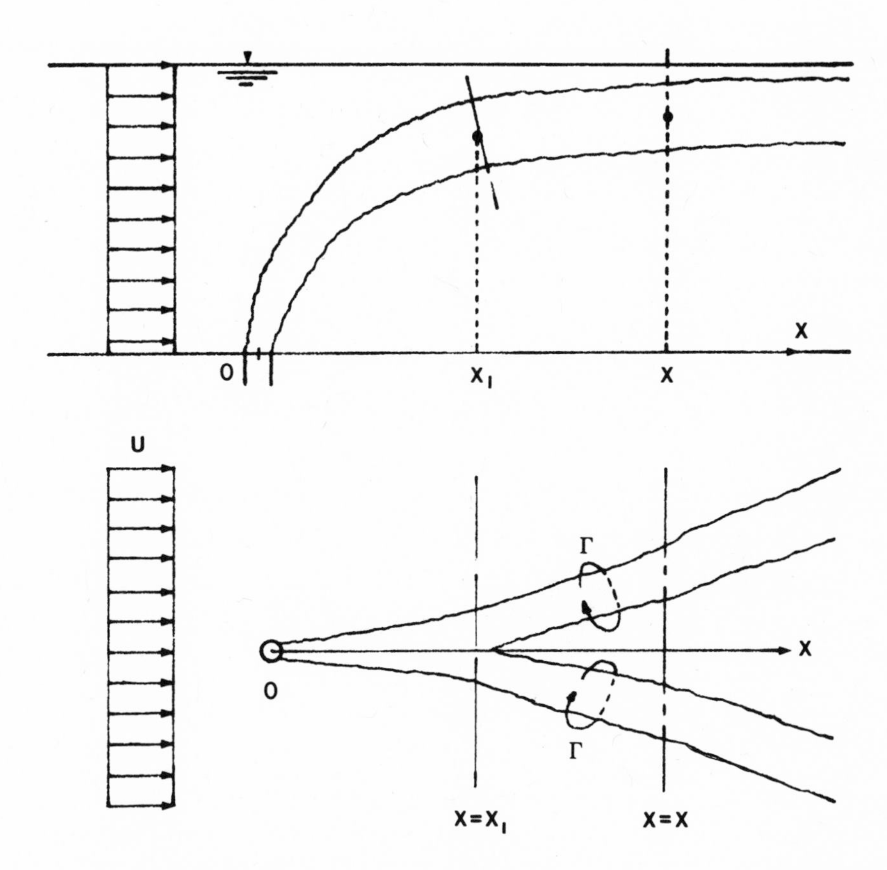

Fig. 5. The bifurcated plume with reference to the horizontal x-axis. Here U is the cross stream velocity, 0 is the spill site, $x = x_1$ is the position where bifurcation starts and Γ is the circulation associated with the vortex pair.

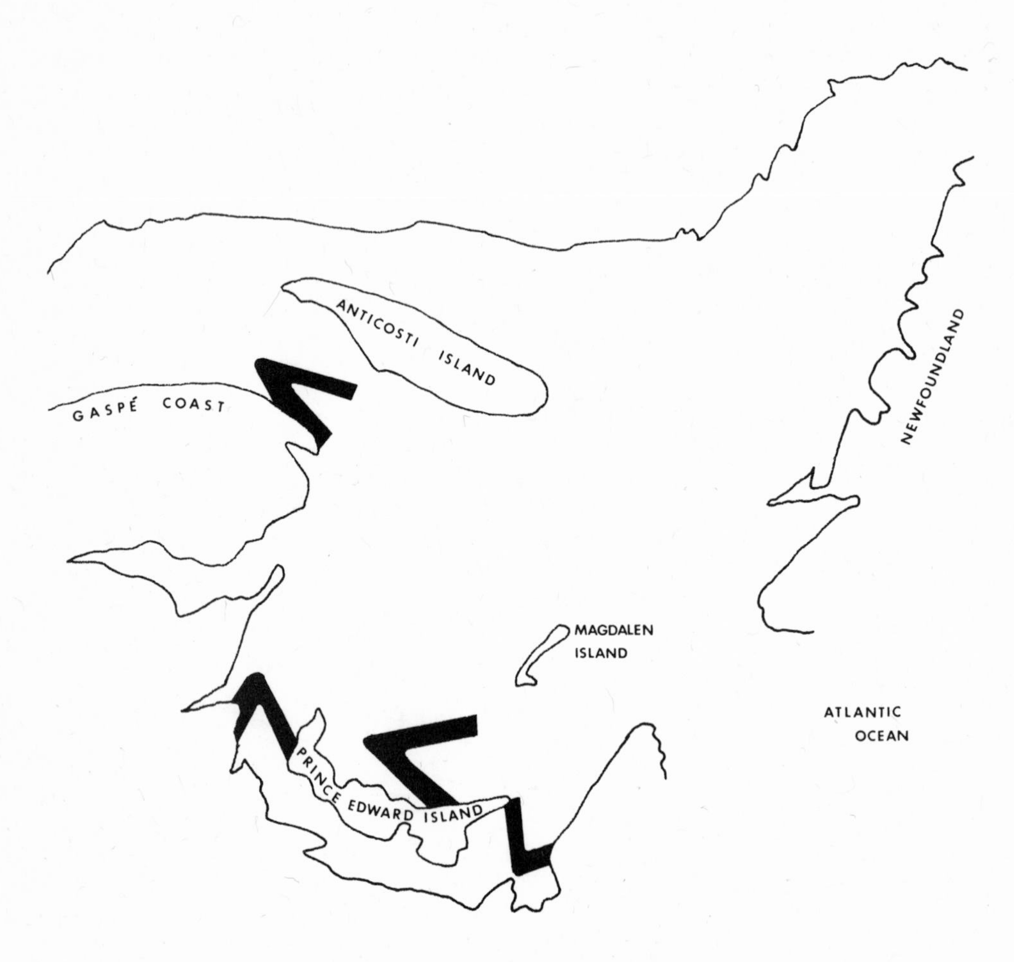

Fig. 6. Simulation of bifurcated plume at four different sites in the Gulf of St. Lawrence.

part is less buoyant which is left obliquely below the inner part. In this process, a vortex pair develops. As the vortex pair rises towards the surface, the free surface produces an image effect (see Lamb, 1945) and the vortex pair which has been combined in a plume separates laterally and then the plume bifurcates. For details on buoyant plumes, see Scorer (1959; Slawson and Csanady, 1967; Khandekar and Murty, 1975 and Murty and Khandekar, 1973).

The angle θ between the horizontal x-axis and each arm of the bifurcated plume is:

$$\tan\theta = \frac{\frac{1}{4\sqrt{2}}\left(\frac{D}{h}\right)^{\frac{1}{2}}\left(\frac{U_0}{U}\right)^{3/2}}{F_0} \tag{7}$$

where D is the diameter of the orifice through which the oil is discharged, U_0 is the velocity of discharge at the orifice, h is the depth below the free surface at which the plume develops a pair of vortices (see Hayashi for details) and F_0 is a densimetric Froude number given by:

$$F_0 = \frac{U_0}{\sqrt{\frac{\Delta\rho}{\rho_w} gD}} \tag{8}$$

where g is gravity, ρ_w is the density of sea water at the location of the spill and $\Delta\rho$ is the difference in the densities of sea water and the particular oil involved in the spill.

Figure 6 shows some simulated bifurcated plumes in the Gulf of St. Lawrence.

ACKNOWLEDGEMENTS

I wish to express my appreciation to Margaret Johnstone for typing the manuscript.

REFERENCES

Assaf, G., R. Gerard and A.L. Gordon. 1971. Some mechanisms of oceanic mixing revealed in aerial photographs. J. Geophys. Res. 76(27): 6550-6572.

Barber, F.G. 1970. Operation oil: some aspects of reconnaissance. In Report of the task force operation-oil. (Clean-up of the "Arrow" oil spill in Chedabucto Bay) to the Minister of Transport 3: 35-68.

Blokker, P.C. 1964. Spreading and evaporation of petroleum products on water. Proc. 4th Int. Harbor Conf., Antwerp, Belgium pp. 911-919.

Fay, J.A. 1969. The spread of oil slicks on a calm sea. In Oil on the sea. (ed.) P. Hoult, Plenum Press 53-79 p.

Hayashi, T. 1971. Turbulent buoyant jets of effluent discharged vertically upward from an orifice in a cross-current in the ocean. Proc. 14th Congress, Int. Assoc. Hydr. Res., Paris, Aug. 29-Sept. 3, 1: 157-165.

Khandekar, M.L. and T.S. Murty. 1975. A note on bifurcation of buoyant bent-over chimney plumes. Atmos. Environ. 9: 759-762.

Lamb, H. 1945. Hydrodynamics.

Longuet-Higgins, M.S. and R.W. Stewart. 1962. Radiation stress and mass transport in gravity waves, with application to surf-beats. J. Fluid Mech. 13: 481-504.

Murty, T.S. and M.L. Khandekar. 1973. Simulation of movement of oil slicks in the Strait of Georgia using simple atmosphere and ocean dynamics. Proc. Joint Conf. on Prevention and Control of Oil Spills, Washington, D.C., July, 1973, 541-546 p.

Murty, T.S. 1974. Remote sensing and numerical circulation studies in ice-covered waters. In Priorities for oceanographic remote sensing. (ed.) J.F.R. Gower, Ocean and Aquatic Affairs, Victoria, B.C. 37-39 p.

Scorer, R.S. 1959. The behavior of chimney plumes. Int. J. Air Pollution 1: 198-220.

Simons, T.J., K. Beal, G.S. Beal, A. El-Sharrawi and T.S. Murty. 1975. Operational model for predicting the movement of oil in Canadian navigable waters. Man. Rep. No. 37, Ocean and Aquatic Sci., Dept. Environ., Ottawa, 30 p.

Slawson, P.R. and G.T. Csanady. 1967. On the mean path of buoyant bent-over chimney plumes. J. Fluid Mech. 28: 311-322.

Starr, V.P. 1968. Physics of negative viscosity phenomena. McGraw Hill Book Company, New York, 256 p.

Stewart, R.W. 1963. Some aspects of turbulence in the Arctic. Proc. Arctic Basin Symp., Hershey, Pennsylvania, October, 1962, Pub. by the Arctic Inst. of North America, 122-127 p.

Winkinson, D.L. 1971. Containment of oil slicks in the St. Lawrence River. Rep. LTR-HY-16. Hydr. Lab., Nat. Res. Council, Ottawa, 55 p.

RELATIONSHIP OF VERTICAL TRANSPORT ACROSS THE THERMOCLINE TO OXYGEN AND PHOSPHORUS REGIMES: LAKE ONTARIO AS A PROTOTYPE

William J. Snodgrass

Dept. of Civil and Chemical Engineering, McMaster University, Hamilton, Ontario

ABSTRACT

Relationships of vertical transport to P and O_2 regimes in stratified lakes is examined using a two-box model perspective. For Lake Ontario, the following statements are supported by this work: 1) Estimates of vertical exchange coefficients across the thermocline range from 0.07-0.15 m/day and for the vertical diffusivity coefficient from 1-4.1 m^2/day. 2) Vertical exchange which transports O_2 from the hypolimnion to the epilimnion accounts for 20% of the loss of oxygen from the hypolimnion; the remainder of the loss is due to decomposition. 3) When a phosphorus-oxygen model is used, the following theoretical predictions are made: (a) increased transport across the thermocline causes decreasing concentrations of total phosphorus at the following spring circulation, (b) vertical transport of orthophosphorus to the epilimnion exceeds inputs to the epilimnion from land-based sources, (c) as the rate of vertical exchange increases, the time to reach steady-state due to a change in loading decreases, (d) downward erosion of the thermocline and the rate of vertical exchange have different effects on a simulated bloom. 4) Predictions of the rate of vertical exchange of oxygen agree fairly well with calculations made from observations.

INTRODUCTION

Of many water quality problems, eutrophication of large lakes has dominated the concern of many Canadians and Americans who live adjacent to the Great Lakes. The eutrophication of lake waters is mainly imputable to man (Vollenweider, 1968).

While universal agreement about its definition, causes, and socio-economic consequences does not exist, the detrimental effects of man-induced eutrophication are generally undeniable. Eutrophication is characterized by the excessive growth of algae, detritus accumulation, exhaustion of dissolved oxygen in the bottom waters, fish kills and associated nuisances. Such characteristics threaten to shorten the life span of lakes and to destroy their utility as a water resource. Total phosphorus concentrations and the rate of development of the hypolimnetic oxygen deficit are two parameters which serve as useful indicators of such characteristics of eutrophication and of the trophic level of lakes.

To examine the relationships among these indicators (i.e., total phosphorus concentration, degree of hypolimnetic oxygen deficit), various biochemical and physical properties within a lake, and external lake variables (i.e. hydraulic loading, nutrient loading, mean depth, surface area), it is useful to formulate mathematical deterministic models. These models relate these various indicators and variables using the appropriate transport processes and biochemical transformations. In particular, investigators such as Vollenweider (1968, 1969, 1975), Dillon and Rigler (1975), Imboden (1973, 1974) and Snodgrass and O'Melia (1975) have been concerned with relating the rate of nutrient input to a lake from land-based sources to the attendant water quality of the lake; others have been considered the relationship between vertical transport across the thermocline and attendant water quality, both in small and in large lakes. A few (e.g., Imboden, Snodgrass and O'Melia) have examined both the nutrient input - water quality and the vertical transport - water quality relationships simultaneously.

The objective of this paper is to examine aspects of the relationship between vertical transport across the thermocline and the phosphorus and oxygen regimes typical of large lakes using Lake Ontario as a prototype. For this examination: (1) data describing oxygen and temperature conditions for 1966-1974 have been analyzed and rates of vertical transport of oxygen across the thermocline calculated, (2) theoretical predictions for vertical transport of a phosphorus model, whose predictions concern total phosphorus concentrations have been tested and verified, are presented and (3) predictions for vertical transport of an oxygen model based upon this phosphorus model are compared with observations.

For examining phosphorus and oxygen regimes in a large lake, it is useful to divide the lake into two boxes consisting of an epilimnion overlying the hypolimnion during summer stratification. The hydraulic input and discharge occur respectively to and from the epilimnion; vertical transport across the thermocline allows transport of materials between the two boxes. Vertical transport across the air-water interface allows atmospheric-epilimnetic exchange of matter; transport across the sediment-water interface allows hypolimnetic exchange of matter with the sediments. During the winter, the lake may be described similarly to the summer configuration if inverse stratification is significant or as one box if the lake is well-mixed.

MODELS OF VERTICAL TRANSPORT ACROSS THE THERMOCLINE

Many investigators have examined methods for calculating vertical transport coefficients or have constructed models of transport processes across the thermocline; several have examined the interrelationships between such transport and the concentrations of various chemical species in the epilimnion and hypolimnion. A few pertinent efforts are reviewed here becuase they provide a base for the present work.

Spalding and Svensson (1976) present a framework for grouping models which calculate the thermocline depth and temperature distribution. There are three types: 1) models which emphasize the radiation and neglect the effect of wind (e.g. Huber et al., 1972), 2) models which compare the energy input from wind and the change in potential energy (They conclude that most models developed to date belong in this class; e.g. Turner and Krauss, 1967, whose model is limited due to profile assumptions concerning temperature and mixing processes.), 3) models which use turbulent transport coefficients (the assumptions and difficulties in specifying the exchange coefficient limit the applicability of this model group). The model of Spalding and Svennson, which is of the last group, avoids the difficulty of determining the exchange coefficient by using a turbulence model from which the exchange coefficient is calculated. Their model is not verified due to a lack of adequate data.

To calculate the vertical transport of mass across a thermocline, models are needed which can predict the correct thermal structure and which can estimate the rate of vertical transport. Since, as Spalding and Svensson's analyses indicate, such models are generally unavailable, resort is normally made to using existing temperature-time-depth data for estimating vertical transport rates. This procedure limits the utilization of such estimates for forecasting purposes.

Measurements of vertical transport rates have been made by many investigators, including Mortimer (1942, 1969), Blanton (1973 a,b), Powell and Jassby (1974), O'Melia (1972), Li (1973), Lerman and Stiller (1969), Bella (1970), and Hesslein and Quay (1973). The majority of such calculations are made using Fick's Law which states that the Flux = $(k_{th})(dc/dz)$ where dc/dz is the vertical gradient of heat or concentration of chemical species across the thermocline, and k_{th} is the vertical transport coefficient across the thermocline. But as Hesslein and Quay note, care must be taken in their utilization, especially in small, highly stratified lakes, not directly affected by winds. In such lakes, the value of the real eddy diffusion may be less than the value of the coefficient of molecular heat conductivity (12×10^{-4} cm^2/sec). They advocate utilization of a tracer such as radon, whose molecular diffusion coefficient (0.2×10^{-4} cm^2/sec) is smaller than that for heat. However, if eddy transport processes are that small, transport of individual chemical species is governed by Brownian processes and is a function of several variables, including size of the atom and temperature of the fluid. Then the correct transport coefficient will have a value equal to that of the molecular diffusivity coefficient of the chemical species of concern or the molecular heat conductivity coefficient in the case of heat, not that of the tracer. Only when turbulence is sufficiently large that the calculated transport coefficient for a chemical species is much larger than the molecular diffusivity coefficient of that species does the transport rate of chemical species approach that of heat.

Several investigators have sought relationships between vertical transport processes and morphometric parameters of lakes. Analysis of Mortimer's (1942) data for the mean vertical eddy diffusion coefficients (k_h, m^2/day) in the hypolimnetic waters of several lakes by Snodgrass and O'Melia (1975) indicates that $k_h = 0.0142\ \bar{Z}^{1.49}$ ($\bar{Z}$, m; correlation coefficient = 0.952). Mortimer's data is based upon both thermal and chemical data. Mortimer expresses the results qualitatively as "The 'average intensity' of turbulent stirring in the hypolimnia of lakes is generally positively correlated with lake dimensions, including depth". Blanton (1973) examined the vertical entrainment of hypolimnetic waters into the epilimnia of lakes and concluded that the greater the mean depth, the greater the entrainment. Analysis of Blanton's data for vertical diffusion coefficients (k) within a region probably characteristic of conditions just above the thermocline (Snodgrass and O'Melia, 1975) indicate that $k = 0.0016\ \bar{Z}^{1.58}$ (k, m^2/day; $\bar{Z}$, m; correlation coefficient = 0.941) for lakes whose mean depth range from 3.2-740 m.

Based upon Blanton's (1973) observations that the mean static stability (sec^{-2}) across a thermocline varies inversely as the mean depth of a lake (i.e., that deep lakes may be less able to resist mixing by fluid turbulence) and Ozmidov's evidence in oceanic waters that the vertical exchange coefficient (e.g., m^2/day) decreases with increasing density gradient, Snodgrass and O'Melia (1975) sought a relationship between vertical transport coefficients across the thermocline (K_{th}, m^2/day) and lake mean depth ($\bar{Z}$, m). Their analysis for 14 lakes whose mean depth ranges from 4.4-740 m indicate that k_{th} = 0.0068 $\bar{z}^{1.12}$ (correlation coefficient = 0.924). Such empirical models provide one approach for examining the effects of vertical transport upon the chemical regimes of several lakes of different mean depth.

For incorporation of the effects of vertical transport into lake models, five works are of note, Rainey (1967), Sweers (1969 a,b, 1970), Stauffer and Lee (1973), Burns and Ross (1972), and Burns (1976 a,b). Rainey (1967) used a one-box model to predict a lake's temporal response to removal of a conservative material for the Great Lakes. The lake was considered as well-mixed year round; stratification was not considered. Sweers (1969 a,b) modified Rainey's model by considering that summer stratification (four months) may be described by a two-box model (epilimnion overlying hypolimnion) connected by appropriate boundary conditions to a one box model for the winter circulation period (eight months). During summer stratification, inputs to and outputs from the lake are assumed to occur only to and from the epilimnion, i.e., the thermocline is an impregnable seal through which no transport occurs. Applied to Lake Ontario for a conservative substance, Sweers found that summer stratification had a negligible effect on the removal time of conservative substances (e.g., compared to Rainey's model). This conclusion is probably valid for time-scales of the order of years in lakes whose hydraulic detention time is much larger than the time-scale of summer stratification; it is probably invalid for lakes whose detention time is of the order of summer stratification or for modelling objectives whose time-scale is of the order of months.

Assumptions that the thermocline may be considered as a seal were justified by Sweer's (1970) development of a quasi-Lagrangian technique for estimating of a vertical diffusion coefficient across the thermocline. (The heat transport equation is transformed onto a quasi-Lagrangian coordinate system whose origin moves down with the mean depth of the thermocline; the eddy diffusivity coefficient is calculated as a function of the lake mean rate of downward displacement of the thermocline and its intensity). Sweer's calculations show that while seasonal averages of the vertical diffusivity of heat are much higher than those of the well developed thermocline (e.g., during quiet weather conditions and during the heating season), the values are much smaller than values for the epilimnion and hence the thermocline effectively acts as a diffusion floor (or transport barrier) throughout most of the summer.

In contrast to Sweer's assumption, the work of Stauffer and Lee (1973), Burns and Ross (1972) and Burns (1976 a,b) indicate that transport across the thermocline is significant. Stauffer and Lee examined the relationships of upward vertical transport across the thermocline of soluble reactive phosphorus into the epilimnion and the changes in mass of total epilimnetic phosphorus and chlorophyll. Estimates of the vertical transport rate were made by monitoring the increase of temperature over time and using Fick's First Law. Conceptually, their calculation procedures allow separation of the net transport into two mechanisms: a vertical two-way water exchange analogous to a turbulent eddy diffusivity mechanism (expressed mathematically as the transport of a certain volume of epilimnetic water, including its heat and mass into the mesolimnion and a back transport of an equal volume of mesolimnetic water into the epilimnion), and a one-way upward transport of the upper part of the mesolimnion into the epilimnion (also called entrainment or erosion of the mesolimnion into the epilimnion). However, their actual calculations incorporate transport resulting from upwelling and transport involving entrainment of colder water because of changes in the Richardson number accompanying the currents set up by Langmuir circulation helixes into net one-way transport for each time period between temperature observations.

Stauffer and Lee found that their estimates of vertical transport of soluble reactive phosphorus for Lake Mendota were somewhat larger than but generally showed good agreement with increases in total epilimnetic phosphorus for five time periods (The lower nature of the actual total phosphorus increase is attributed to losses due to settling from the epilimnion). In particular, their data show that on two separate occasions, a large increase in the mass of chlorophyll corresponds to a significant downward migration of the thermocline. Since reactive phosphorus (assumed to be biologically available) is the main form of P transported into the epilimnion due to entrainment, significant new algae growth can occur. This results in large increases in the total chlorophyll content. An increase in volume of the epilimnion accounts for some of the increase in mass of chlorophyll, but is insufficient to explain to the total increase. In addition, Stauffer and Lee assert that most of the epilimnetic entrainment occurs during the passage of cold fronts and storms which have windy characteristics sufficient to cause entrainment events.

While it is difficult to establish a complete cause-effect relationship between vertical nutrient transport and subsequent increases in chlorophyll mass, Stauffer and Lee present arguments concerning algae growth requirements to justify their conclusion that periodic nutrient enrichment of the epilimnion due to mesolimnetic entrainment provides for maintenance of high summer chlorophyll levels. While their data is not conclusive (settling rates from the epilimnion were not measured; data is presented to show that rainfall and runoff inputs during the May-October period are negligible — total stream inputs to the epilimnion are estimated to be equal to inputs from the mesolimnion during the passage of one storm front), their investigation does provide substantive support for the hypothesis that vertical transport and in particular mesolimnetic entrainment are a dominant source of inputs of phosphorus into the epilimnion.

In contrast to Stauffer and Lee, Burns and Ross (1972) and Burns (1976 a,b) developed models to describe transport for a lake (Central Basin of Lake Erie, 1970) in which entrainment can occur either upward or downward. Three processes are considered to be important: 1) erosion of a layer of the mesolimnion downward into the hypolimnion (hypolimnetic volume increase, 2) erosion of a layer of the hypolimnion upwards into the mesolimnion (hypolimnetic volume decrease), and 3) simultaneous downwelling of part of the mesolimnion and epilimnion into the hypolimnion and upwelling of part of the hypolimnion into the mesolimnion or the epilimnion elsewhere in the basin (no volume change in hypolimnion). The Mesolimnion Erosion Model of Burns and Ross considered only the first two processes; the Mesolimnion Exchange Model of Burns (1976 a) considered all three processes. Changes in the volumes and heat content of the epilimnion, mesolimnion and hypolimnion between successive profiting surveys and framed in a heat budget context permitted calculations of the direction and amount of mass (e.g., oxygen, phosphorus) transported across the thermocline. Burns (1976 a) found that the exchange process comprised about 12% of the hypolimentic oxygen depleted during 1970, reaching 19% during one period. Burns (1976 b) found that vertical transport from the hypolimnion into the epilimnion was the major source of in the epilimnetic budget of soluble reactive silica (70% of total inputs to the epilimnion) but of lesser importance for in the epilimnetic budget of soluble inorganic nitrogen (NH_3 + NO_3^- + NO_2^-; 35% of total inputs). For total phosphorus, the dominant mechanism was sedimentation from the epilimnion; hypolimnetic transport of TP into the epilimnion was quite small. This is expected since the dominant upward transport mechanism would be of soluble reactive phosphorus - use of TP masks this mechanism. Burn's (1976 b) data indicate that vertical transport across the thermocline accounts for approximately 35% of total phosphorus inputs in the Central Basin's epilimnion during the 1970 stratification.

CALCULATIONS OF OXYGEN TRANSPORT ACROSS THE THERMOCLINE

For 1966-1974, the Canada Centre for Inland Waters has sampled Lake Ontario at various depths of a 60-90 station grid on approximately a monthly basis (April-November). The data for each cruise were summarized by this investigator into lake-wide average values as a function of depth for temperature, oxygen, and phosphorus by calculating a mean value for a particular horizontal layer for each parameter.

For each vertical temperature profile, the location of the thermocline was graphically determined (the thermocline was defined as the depth of maximum temperature gradient). The depth of the thermocline is plotted as a function of time for a few typical years in Figure 1 and the rate of deepening of the thermocline (calculated using the method of least-squares) is shown in Table 1. These data indicate that 1967 and 1968 were relatively calm years while the remainder were more active. However, the thermocline in 1968 was consistently 8 meters deeper than 1967 (see Fig. 1). This indicates that vertical transport in the late spring of 1968 was probably high, providing for thermocline formation at a deeper level.

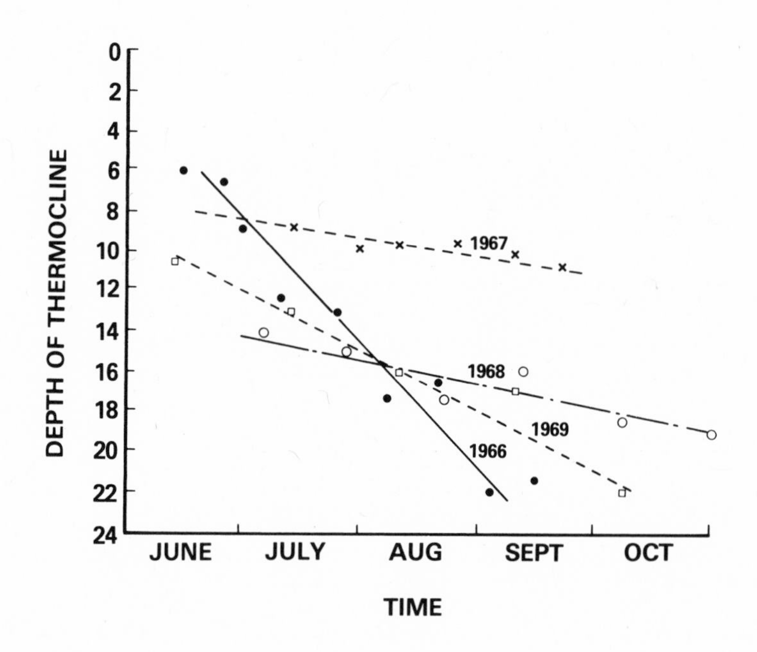

FIG. 1

Depth of Epilimnion ($\overline{Z_e}$) as a Function of Time.

TABLE 1
Rate of Increase of Depth of Thermocline

Year*	Rate (m/month)
1966	5.5
1967	0.8
1968	1.1
1969	2.8
1970	3.7
1972	5.8

*Data for 1971,73, and 74 is insufficient to allow calculation.

TABLE 2
Coefficients Describing the Rate of Vertical Transport

Year	Vertical Exchange Coefficient (m/day)	Vertical Diffusivity Coefficient (m^2/day)
1966	0.14	4.1
1967	0.12	1.4
1968	0.07	1.0
1969	0.12	2.9
1970	0.13	2.6
1972	0.15	3.4

Vertical exchange coefficients and vertical diffusivity diffusion coefficients estimated from these data, are shown in Table 2. The method of estimating the coefficients is shown in Table 3. Based upon Fick's First Law, these vertical transport coefficients integrate the effects of molecular diffusion, eddy diffusion, internal waves, seiches, standing waves, hypolimnetic entrainment into the epilimnion and leakage around the sides into a net transport process. The vertical diffusivity coefficient is analogous to a diffusion coefficient; the vertical exchange coefficient is a mass transfer coefficient which conceptually is equal to the vertical diffusivity coefficient divided by the thickness of the thermocline (i.e., a thickness over which the concentration gradient exists). The coefficients are calculated as average values for the whole stratification period and are representative of the average depth of the thermocline over the stratification period.

These values suggest that the vertical exchange coefficient can vary by a factor of two while the vertical diffusivity coefficient varies by a somewhat greater amount. However, the values are considered only as preliminary estimates until more precise estimates are obtained by using volume-weighted

TABLE 3
Calculation of the Vertical Transport Coefficients

Vertical Exchange Coefficient ($\hat{k}_{th}$ m/day)

$$\hat{k}_{th} = \frac{V_h}{A_{th}} \left[\frac{T_h(t) - T_h(0)}{\Delta t(\bar{T}_e - \bar{T}_h)}\right]$$

Vertical Diffusivity Coefficient (k_{th} m^2/day)

$$k_{th} = \frac{V_z}{A_z} \frac{(T_z(t) - T_z(0)) \cdot \Delta z}{\Delta t(\bar{T}_{z-1} - \bar{T}_{z+1})}$$

Symbols

V_h	average volume of hypolimnion over time t(m^3)
V_z	average volume of lake below depth z over Δt (m^3)
A_{th}	average interfacial area of thermocline over time t (m^3)
A_z	average interfacial area at depth z over Δt (m^2)
Δt	time period (days)
$T_h(t)$	temperature of hypolimnion at time t (^oC)
$T_h(0)$	temperature of hypolimnion at time 0 (^oC)
$T_z(t)$	average temperature of water below depth z at time t(^oC): Note, this is calculated as $\Sigma V_i T_i / \Sigma V_i$ where T_i is the average temperature in a layer of water of volume V_i
$T_z(0)$	average temperature of water below z at time 0
$\bar{T}_e$	average temperature of epilimnion over Δt (^oC)
$\bar{T}_h$	average temperature of hypolimnion over Δt (^oC)
$\bar{T}_{z-1}$	average temperature of depth z-1 over Δt (^oC)
$\bar{T}_{z+1}$	average temperature of depth z+1 over Δt (^oC)
Δz	thickness of layer of calculation ($\Delta z = \lvert z+1 \rvert - \lvert z \rvert$)

calculations of heat content. The values for vertical diffusivity are somewhat higher than those estimated by Sweers (1970) (1.9 and 0.4 m^2/day for 1966 and 1967 respectively). Sweers' method (calculating diffusivity coefficients in the region of the thermocline and averaging the values for the adjacent areas) uses a shorter time period and a 5 m layer thickness (rather than a 2 m thickness used here). The longer period was used here so that the time period for calculating both the vertical exchange and vertical diffusivity coefficients would be the same and so that the complete four-month stratification period would be described.

Based upon these lake-wide averages as a function of depth, volume-weighted oxygen concentrations were calculated for the epilimnion and hypolimnion. (For example, the dissolved oxygen concentration in the hypolimnion for any cruise is equal to $(\Sigma C_i V_i)/V_h$ where C_i is the lake-wide average concentration of dissolved oxygen calculated for the layer whose volume is V_i and $V_h = \Sigma V_i$). The temporal variation of oxygen as a function of time is shown for 1970 in Figure 2. Based upon similar profiles for other years, the average rate of loss of hypolimnetic oxygen is calculated and shown in Table 4.

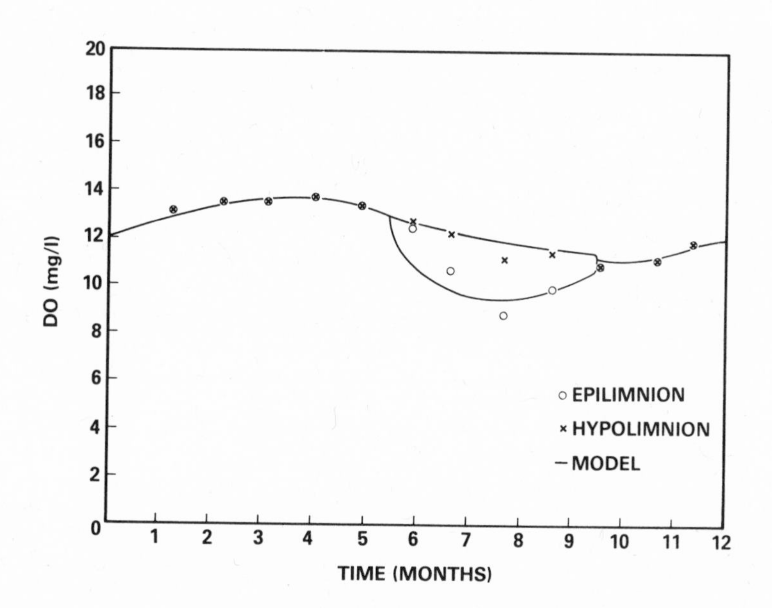

FIG. 2
Dissolved Oxygen as a Function of Time in Lake Ontario (1970).

TABLE 4

Observed D.O. Fluxes Associated with the Hypolimnion

Year	Total Hypolimnetic Oxygen Demand ($gm/m^2/day$)	Vertical Exchange	Decomposition
1966	0.49	57%	43%
1967	1.3	15%	85%
1968	0.91	7%	93%
1969	1.3	13%	87%
1970	1.4	15%	85%
1972	1.5	12%	88%

To determine the relationship of vertical exchange and decomposition to the hypolimnetic depletion of oxygen, estimates of vertical transport of oxygen from the hypolimnion are made and the differences are ascribed to decomposition. As is indicated in Figure 2, the hypolimnetic concentration of oxygen is greater than the epilimnetic concentration of oxygen during stratification. Hence, vertical transport of oxygen occurs from the hypolimnion to the epilimnion and represents a loss of oxygen from the hypolimnion. For any year, the flux due to vertical transport is calculated as being equal to the product of the vertical exchange coefficient (m/day), and the concentration difference of dissolved oxygen between the epilimnion and hypolimnion (gm/m^3). Since the concentration gradient changes during stratification, estimates of the vertical exchange flux are made at different times and then weighted according to the length of time that each flux estimate occurs to give a seasonal estimate of the flux. These estimates for vertical exchange are expressed as a percent of the total hypolimnetic loss and shown in Table 4. The remaining loss of hypolimnetic oxygen is assumed to be due to decomposition (water column plus sediments).

Two observations are noteable: (i) Vertical exchange acts to account for an average of 20% of the total hypolimnetic loss of oxygen, demonstrating that the thermocline is only a partial seal to transport; (ii) The net transport of oxygen occurs from the hypolimnion to the epilimnion during stratification.

THEORETICAL PREDICTIONS CONCERNING PHOSPHORUS

This section examines, from a model perspective, the relationship of vertical transport across the thermocline to epilimnetic stocks of phosphorus, nutrient inputs from land-based sources, and time for a lake to respond to changes in nutrient inputs.

SUMMER STRATIFICATION

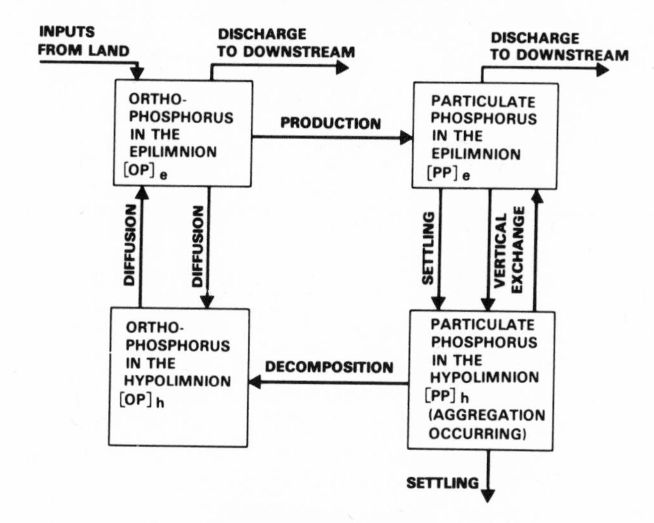

WINTER CIRCULATION

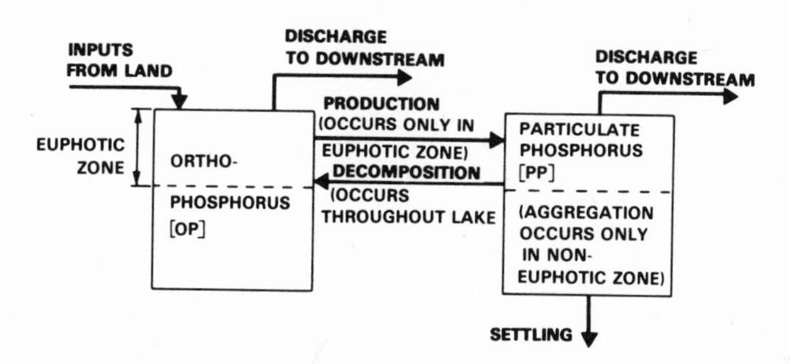

FIG. 3
Compartments and Mass Flows of Phosphorus Model.

A predictive model for phosphorus concentrations in lakes has been developed and verified (Snodgrass, 1974; Snodgrass and O'Melia, 1975). This model is presented schematically in Figure 3. Summer stratification is described by two boxes, winter circulation is described by one box. All phosphorus species are represented by only two forms, soluble orthophosphorus (OP) and particulate phosphorus (PP). Reactions utilized are represented by the appropriate arrows in Figure 3. Equations describing the model and coefficient values are the same as presented in Snodgrass and O'Melia (1975), except that the length of the stratification period for Lake Ontario is taken as four months. Model predictions concerning the effects of vertical transport upon various lake parameters are now presented. Since these predictions are not verified due to a lack of appropriate field data, they are considered as theoretical predictions.

The effects of increased vertical transport are shown in Figure 4. The stable-cycle profile represents the model prediction of the annual phosphorus cycle for the condition where a lake is at 'steady-state' with a particular rate of vertical transport. The values for the vertical exchange coefficient have a range which is greater than those calculated (see Table 2) and is assumed to represent the range between very calm conditions and fairly stormy conditions. As the rate of vertical exchange increases, the concentration of total phosphorus at the following spring circulation is predicted to decrease. For consistently calm conditions, more phosphorus builds up in the hypolimnion during summer stratification than for more stormy conditions; hence, under calm conditions, a larger mass of phosphorus is found in the lake at fall overturn and this larger mass is observed at the end of spring circulation.

For these same mixing conditions, model predictions of the net loading of soluble orthophosphorus to the epilimnion due to

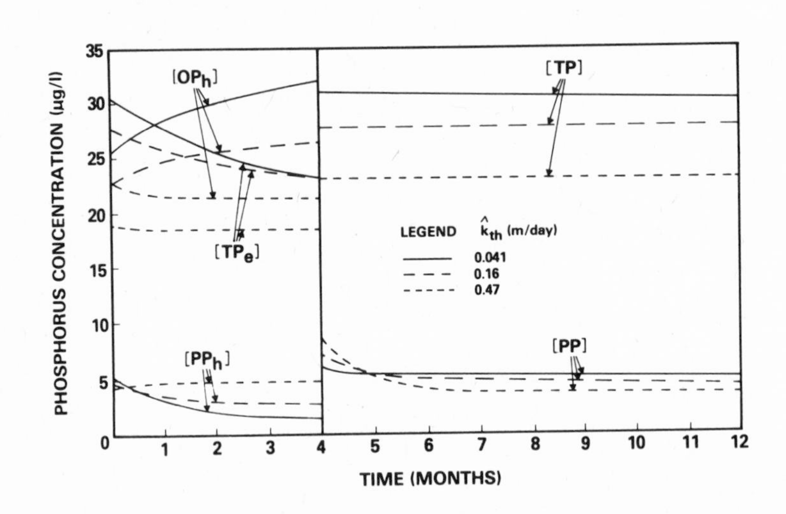

FIG. 4

Effect of Different Rates of Vertical Exchange ($\hat{k}_{th}$) on the Annual Phosphorus Concentrations (Coefficients as in Snodgrass and O'Melia, 1975).

vertical exchange are shown in Table 5. The loading to the epilimnion from land-based sources is 1.8 mg/m^2-day. For a vertical exchange rate of 0.04 m/day, loadings from the land and due to vertical exchange are in balance. For higher rates, the loading due to vertical exchange is greater than the loading from land. The calculated vertical exchange rate is generally greater than 0.04 m/day. This indicates that the loading from the hypolimnion will be as important or more important than that from land based sources during summer stratification.

To simulate the relationship of vertical transport to an algae bloom, two cases are considered: (i) the rate of vertical exchange is increased by an order of magnitude for a week, and (ii) the depth of the thermocline is increased by 5m over a week. The former simulates the effects of increased vertical diffusivity, the latter simulates the effects of entrainment of hypolimnetic water into the epilimnion. Model predictions are shown in Figure 5.

An increase of the vertical exchange coefficient ($\hat{k}_{th}$) acts to equalize the concentrations of soluble orthophosphorus ([OP]) and particulate phosphorus ([PP]) between the epilimnion and hypolimnion. Accordingly, the epilimnetic [OP] increases but the epilimnetic [PP] decreases; the concentration of total epilimnetic phosphorus increases slightly. After mixing ceases (day 67), an increase in biomass will occur but only approximately to the initial concentration of day 60 (at steady-state with a given mixing rate, the ratio of $[PP]_e$ and $[TP]_e$ is approximately constant). An increase in the depth of the thermocline acts to increase epilimnetic soluble and total phosphorus and to decrease particulate phosphorus. The hypolimnetic concentrations of soluble and total phosphorus are greater than the respective concentrations in the epilimnion at day 60; the hypolimnetic concentration of particulate phosphorus is smaller than that in the epilimnion. Entrainment of water containing low concentration of particulate phosphorus acts to dilute the concentration of epilimnetic particulate phosphorus since even though entrainment causes the mass of particulate phosphorus in the epilimnion to increase, the volume of the epilimnion increases by a greater amount and the mass per unit volume (concentration) decreases. Similar arguments (e.g., the mass of soluble phosphorus increases by a greater amount then the volume of the epilimnion) support an increase in epilimnetic soluble and total phosphorus concentration. But production should act to increase the particulate phosphorus concentration. Hence, in this case, dilution of particulate phosphorus by hypolimnetic water is greater than production.

This simulation demonstrates that the two transport mechanisms have different effects upon hypolimnetic species —

TABLE 5
Predictions of Soluble Orthophosphorus Inputs
To the Epilimnion due to Vertical Exchange

Vertical Exchange Coefficient (m/day)	Loading (mg/m^2-day)
0.04	1.7
0.16	5
0.47	10

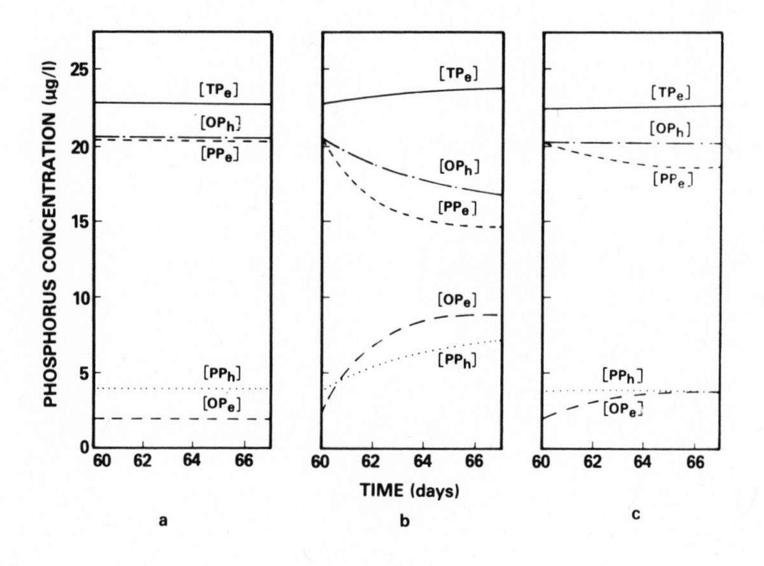

FIG. 5
Stable Cycle Profile Predicted by Phosphorus Model for Day 60 to 67 for Three Entrainment and Vertical Exchange Conditions (Coefficients as in Snodgrass and O'Melia, 1975, except p_e=0.2 day^{-1}).

5(a) Utilizing Standard Constant Coefficients ($\hat{k}_{th}$=0.44 m/day, $\overline{Z_e}$=17 m)

5(b) Allowing $\hat{k}_{th}$ to be an Order of Magnitude Higher than Standard Conditions ($\hat{k}_{th}$=4.4 m/day, $\overline{Z_e}$=17 m)

5(c) Allowing $\overline{Z_e}$ to increase at Rate of 0.73 m/day ($\hat{k}_{th}$=0.44 m/day) entrainment has an inconsequential effect. But the simulation suggests that the epilimnetic concentration of total phosphorus is not expected to substantively increase. For Lake Ontario, this is true since predictions of epilimnetic and hypolimnetic concentrations of total phosphorus are approximately equal. For lakes in which hypolimnetic concentrations are substantively higher than those in the epilimnion (e.g. Stauffer and Lee, 1973), the epilimnetic concentrations of both soluble and total phosphorus will increase, providing a pool of soluble phosphorus available for rapid incorporation into biomass (e.g. a bloom). Such rapid incorporation will occur if the rate of production is more rapid than the effects of mixing or upon ceasation of the transport event.

The effect of an increased rate of vertical transport on the time for the lake to reach steady-state is shown in Table 6. Here the time to reach 95% of the 'steady-state' concentration, $\tau_{0.95}$ is used as an estimate of the time required for most of the change in water quality due to a change in inputs rate of phosphorus to be observed. The model predicts that the time to reach steady state decreases as the rate of vertical transport increases.

MODEL PREDICTIONS OF OXYGEN TRANSPORT ACROSS THE THERMOCLINE

The compartments selected for the oxygen model are shown in Figure 6. The equations for the oxygen model are shown in Table 7. The same coefficient values as Snodgrass and O'Melia (1975) are utilized for the phosphorus model, except that the calculations of a thermocline depth as a function of time during stratification and the values for $\hat{k}_{th}$ for different years (see Tables 1,2) are used. The value selected for p_e(0.2 day^{-1}) has an insignificant effect on model predictions of [TP] and [PP] decomposition (Snodgrass, 1974) but predicts a more representative value for [OP_e]. For the oxygen model, the value of R selected is 138 μM O_2/μM P (utilizing the Redfield stoichiometric relationship; Stumm and Morgan, 1970), SOD is 0.1 gm/m^2-day, k_{as} is 2 m/day and k_{aw} is 7.5 m/day. The values for k_{as} and k_{aw} are determined by calibrating the oxygen model using 1966 and 1969 data for Lake Ontario after selection of the value for SOD.

For verification, the model's predictions for oxygen are compared with observations for 1970 in Figure 2. Agreement is good. This provides some verification for the model.

A second method for model verification is to compare model predictions of the vertical flux during summer stratification with observations. The comparison (see Table 8) shows that model predictions are slightly higher but that model predictions agree quite well with observations. This provides additional verification of the model.

TABLE 6
Time for Lake to Reach 95% of 'Steady-State' Concentrations

$\hat{k}_{th}$ (m/day)	$\tau_{0.95}$ (yrs)
0.04	13.5
0.22	11.6
0.45	10.1
0.67	9.2

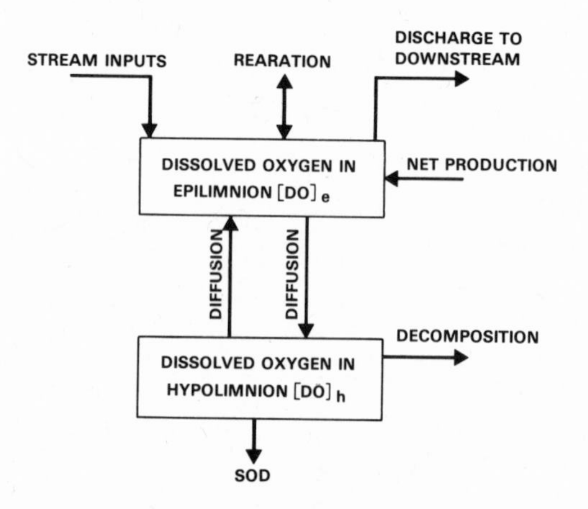

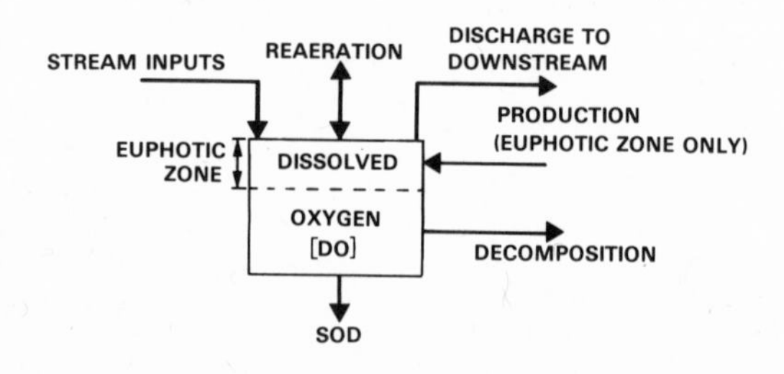

FIG. 6
Compartments and Mass Flows of Oxygen Model

TABLE 7
Equations of Oxygen Model

Summer Stratification

$$V_e \frac{d[DO_e]}{dt} = \Sigma Q_j[DO_j] - Q[DO_e] + k_{as}A_a([DO_s] - [DO_e])$$

Net rate of change; input: loading from land-based sources; output: hydraulic discharge; atmospheric reaeration

$$- k_{th} A_{th} ([DO_e] - [DO_h]) - R\, p_e\, V_e\, [OP_e]$$

output: vertical exchange; input: net production

$$V_h \frac{D[DO_h]}{dt} = k_{th} A_{th} ([DO_e] - [DO_h]) - R\, D_h\, V_h\, [PP_h]$$

Net rate of change; input: vertical exchange; loss: decomposition

$$- SOD\, A_s$$

loss: sediment oxygen demand

Winter Circulation

$$V \frac{d[DO\ ]}{dt} = \Sigma Q_j[DO_j] - Q[DO] + k_{as}\, A_a([DO_s] - [DO])$$

Net rate of change; input: loading from land based sources; output: hydraulic discharge; atmospheric reaeration

$$+ R\, p_{eu}\, V_{eu}\, [OP] - R\, d_h\, V_h\, [PP_h] - SOD\, A_s$$

input: production in euphotic zone; loss: decomposition; loss: sediment oxygen demand

TABLE 8
Vertical Exchange Flux of Dissolved Oxygen for Lake Ontario

Year	Observed Flux (gm/m^2-day	Predicted Flux gm/m^2-day
1966	0.28	0.28
1967	0.19	0.22
1968	0.06	0.15
1969	0.17	0.22
1970	0.22	0.23
1972	0.18	0.26

SUMMARY AND CONCLUSIONS

Models for vertical transport across the thermocline of stratified lakes and calculations relating vertical transport to chemical regimes in a few lakes are reviewed. Similar relationships are examined for Lake Ontario using calculations from field data and using a phosphorus and oxygen model.

To evaluate transport characteristics across the thermocline during summer stratification in Lake Ontario, estimates of the rate of deepening of the thermocline and of the vertical transport coefficients have been made for the period 1966-74 utilizing temperature data. These estimates show that 1967 and 1968 had relatively low vertical transport rates, but that 1966 and 1972 were particularly turbulent. These estimates were applied to oxygen data and to a phosphorus and oxygen model for the lake.

The following effects of vertical transport across the thermocline on the phosphorus regime are predicted by the model: (a) increasing rates of transport act to cause decreasing spring concentrations; (b) vertical transport of soluble orthophophosphorus to the epilimnion exceeds inputs from land-based sources for many mixing conditions; (c) as the rate of vertical exchange increases, the time to reach 'steady-state'decreases; (d) downward erosion of the thermocline and increased vertical exchange have different effects on the occurrence of a bloom. These predictions are theoretical, as adequate field data for verification is lacking.

For the oxygen regime, vertical exchange causes transport of oxygen from the hypolimnion to the epilimnion during summer stratification and accounts for 20% (on the average) of the total hypolimnetic loss of oxygen. Model predictions of oxygen concentrations over time show good agreement with observations; model predictions of the vertical flux of oxygen across the thermocline agree fairly well with calculations made from the field data. Accordingly, these two types of agreement (concentrations, fluxes) provide some verification for the oxygen model.

ACKNOWLEDGEMENTS

The cruise data for Lake Ontario was kindly furnished by the Canada Centre for Inland Waters, Burlington, Ontario, Canada. The data synthesis and oxygen modelling for Lake Ontario was supported both by McMaster University and the University of North Carolina at Chapel Hill. The development of the oxygen model will form a part of the Master's Thesis of Mr. R.J. Dalrymple, Environmental Engineering Group, McMaster University. The help of Mrs.S. Gallo, Mr. H.T. Snodgrass, Dr. C.R. O'Melia, Dr. R.Shaw, and Dr. P.J. Dillon for stimulating this writer to prepare this manuscript or for aiding in its preparation is acknowledged.

NOMENCLATURE

A_a	surface area of atmosphere - water interface (L^2)
A_s	surface area of the sediment - water interface (L^2)
A_{th}	horizontal cross-sectional area of the lake at the thermocline (L^2)
d	decomposition rate coefficient for the whole lake (T^{-1})
d_h	decomposition rate coefficient for the hypolimnion (T^{-1})
k_{aw}	atmospheric mass transfer coefficient in winter (L/T)
k_{as}	atmospheric mass transfer coefficient in summer (L/T)
k	diffusivity coefficient in region above thermocline (L^2/T)
k_h	diffusivity coefficient in hypolimnion (L^2/T)
k_{th}	diffusivity coefficient in the thermocline region (L^2/T)
$\hat{k}_{th}$	vertical exchange coefficient in the thermocline region (L/T)
Q	volumetric rate of discharge from the lake (L^3/T)
Q_j	volumetric rate of inflow to a lake from source j (L^3/T)
R	Redfield Stoichiometry factor (M/L^3 of $O_2/M/L^3$ of P)
p_e	net production rate coefficient in the epilimnion (T^{-1})
p_{eu}	production rate coefficient in the euphotic zone, circulation zone (T^{-1})
SOD	sediment oxygen demand ($M/L^2/T$)
t	time (T)

V	volume of the whole lake (L^3)
V_e	volume of the epilimnion (L^3)
V_h	volume of the hypolimnion (L^3)
V_{eu}	volume of the euphotic zone during winter circulation (L^3)
$\overline{Z}$	mean depth of the lake (L^3)

Concentrations

[DO]	concentration of dissolved oxygen (M/L^3)
[OP]	concentration of soluble orthophosphorus (M/L^3)
[PP]	concentration of particulate phosphorus (M/L^3)
[TP]	concentration of total phosphorus (M/L^3)

Note: subscript e denotes the epilimnion; subscript h denotes the hypolimnion; subscript j denotes source j

Greek Letters

$\tau_{0.95}$ time for lake to reach 95% of 'steady-state' concentration

Note: Symbols particular to Table 5 are shown only there.

REFERENCES

Bella, D.A., 1970. Dissolved Oxygen Variation in Stratified Lakes. Proc. ASCE, J. SED, 96 (SA5), 1129-1146.

Blanton, J.O., 1973. Vertical Entrainment into the Epilimnia of a Stratified Lake. Limnol. Oceanogr., 18, 697-704.

Blanton, J.O., 1973. Rates of Vertical Entrainment in Stratified Lakes. International Association of Hydrological Sciences, Pub. No. 109, 301-305.

Burns, N.M., Ross, C., 1972. Oxygen-Nutrient Relationships Within the Central Basin of Lake Erie. In N.M. Burns and C. Ross (ed.) "Project Hypo". Can. Centre for Inland Waters Paper 6, 85-119.

Burns, N.M., 1976a. Oxygen Depletion in the Central and Eastern Basins of Lake Erie, 1970. J. Fish. Res. Board Can., 33, 512-519.

Burns, N.M., 1976b. Nutrient Budgets for Lake Erie, 1970. J. Fish. Res. Board Can., 33, 520-536.

Dillon, P.J., and Rigler, F.H., 1975. A Simple Method for Predicting the Capacity of a Lake for Development Based on a Lake Trophic Status. J. Fish. Res. Board Canada, 32, 1519-1531.

Hesslein, R., Quay, P., 1973. Vertical Eddy Diffusion in the Thermocline of a Small Stratified Lake. J. Fish. Res. Board Canada, 30, 1491-1500.

Huber, W.C., Harleman, D.R.F., and Ryan, P.J., 1972. Temperature Prediction in Stratified Reservoirs. Proc. Am. Soc. Civ. Engr., J. Hyd. Div., 98, 645-666.

Imboden, D.M., 1973. Limnologische Transport und Nahrstoffmodelle. Schweiz. Z. Hydrol., 35, 29-68.

Imboden, D.M., 1974. Phosphorus Model of Lake Eutrophication. Limnol. Oceanogr., 19, 297-304.

Lerman, A., and Stiller, M., 1969. Vertical Eddy Diffusion in Lake Tiberias. Verh. Int. Verein. Limnol., 17, 323-333.

Li, Yuan-Hui, 1973. Vertical Eddy Diffusion Coefficient in Lake Zurich. Schweiz. Z. Hydrol., 35, 1-7.

Mortimer, C.H., 1942. The Exchange of Dissolved Substances Between Mud and Water in Lakes, III and IV. J. Ecol., 30, 147-201.

Mortimer, C.H., 1969. Physical Factors with Bearing on Eutrophication in Lakes in General and in Large Lakes in Particular. In "Eutrophication: Causes, Consequences, Correctives". National Academy of Sciences, Washington, D.C., 340-368.

O'Melia, C.R., 1972. An Approach to the Modelling of Lakes. Schweiz. Z. Hydrol., 34, 1-33.

Ozmidov, R.V., 1965. On the Turbulent Exchange in a Stably Stratified Ocean. Izv. Bull. Acad. Sci. USSR, Atmospheric and Ocean Physics Ser., 1, 853-860 tr. by Am. Geophys. Union, Washington, D.C., 493-497.

Powell, T., and Jassby, A., 1974. The Estimation of Vertical Eddy Diffusivities Below the Thermocline in Lakes. Water Resour. Res., 10, 191-198.

Rainey, R.H., 1967. Natural Displacement of Pollution from the Great Lakes. Science, 155, 1242-1243.

Snodgrass, W.J., 1974. A Predictive Phosphorus Model for Lakes-Development and Testing. Ph.D. Dissertation, University of North Carolina at Chapel Hill, 309p.

Snodgrass, W.J., O'Melia, C.R., 1975. A Predictive Model for Phosphorus in Lakes. Environ. Sci. and Tech., 10, 937-944.

Spalding, D.B., Svensson, U., 1976. The Development and Erosion of the Thermocline. Imperial College of Science and Technology, London, England. Paper presented at 1976 3 CHM meeting, Dubrovnic, 10pp.

Stauffer, R.E., and Lee, G.F., 1973. The Role of Thermocline Migration in Regulating Algae Blooms in Middlebrooks, E.J., et al, (ed.) "Modelling the Eutrophication Process". Proceedings of Workshop, Div. of Envir. Engr., Utah State Univ., Logan, Utah, 73-82.

Stumm, W., Morgan, J.J., 1970. Acquatic Chemistry. Wiley, 583p.

Sweers, H.E., 1969a. Structure, Dynamics and Chemistry of Lake Ontario. Manuscript Report Series, No. 10, Mar. Sci. Bi., Dept. of Energy, Mines, and Resources, Ottawa, Can.

Sweers, H.E., 1969b. Removal of Contaminants from Lake Ontario by Natural Processes. Proc. 12th Conf. Great Lakes Res., Inter. Assoc. Great Lakes Res., 734-741.

Sweers, H.E., 1970. Vertical Diffusivity Coefficient in a Thermocline. Limnol. Oceanogr., 15, 273-280.

Turner, J.S., Krauss, E.B., 1967. A One-Dimensional Model of the Seasonal Thermocline. Tellus, 19, 88-105.

Vollenweider, R.A., 1968. Scientific Fundamentals of the Eutrophication of Lakes and Flowing Waters, with Particular Reference to Nitrogen and Phosphorus as Factors in Eutrophication. OECD Tech. Rep. DAS/CS1, 68, 27, Paris, 159pp.

Vollenweider, R.A., 1969. Möglichkeiten und Grenzen Elementarer Modelle der Stoffbilanz von Seen. Arch. Hydrobiol., 66, 1-36.

Vollenweider, R.A., 1975. Input-Output Models. Schweiz Z. Hydrol., 37, 53-84.

SURFACE HEAT EXCHANGE AND HYDROTHERMAL ANALYSIS

John Eric Edinger
Edward M. Buchak

J. E. Edinger Associates, Inc.

Abstract

The seven physical mechanisms of surface heat exchange are discussed and their summary as the net rate of surface heat exchange is defined. The algebraic formulation of the net rate of surface heat exchange into an equilibrium temperature and a coefficient of surface heat exchange is presented. The definitions of the equilibrium temperature and coefficient of surface heat exchange lead to the concept of excess temperature resulting from man-made heat sources.

The incorporation of surface heat exchange into hydrothermal analyses is presented in terms of three spatial scales and three spatial dimensions over which heat transport can take place. The relative importance of surface heat exchange to initial discharge mixing, the intermediate scale of advection and dispersion, and large scale whole waterbody processes is formulated and presented. The significance of surface heat exchange in time varying temperature problems is also discussed.

Introduction

Hydrothermal analysis can be defined as the art of combining hydrodynamic principles with heat continuity relationships for the purpose of determining temperature distributions that result from man-made and naturally occurring heat sources and sinks. Because temperature predictions depend on the results of hydrodynamic analyses, a temperature field cannot be predicted in any more detail than a velocity field.

The hydrodynamicist can state succinctly the mathematics of water body modeling using six variables and six space-time equations. The six variables are: The velocity components in each of three coordinate directions, the fluid pressure, a constituent concentration, and the fluid density. Qualitatively, the six equations are the equations of momentum for each of the three coordinate directions, the equation mass continuity, a constituent mass balance, and an equation of state. The mathematical statement is completed by specifying the appropriate initial and boundary values of the six variables.

The complexity of hydrothermal models is often dictated by the significance of surface heat exchange relative to the advective and dispersive transport processes. The relative complexity can be illustrated using simple transport models that are useful approximations to more complex hydrodynamic problems.

Heat Exchange Analysis

Surface heat excnage is one mechanism of water body heat transport that is important in the hydrothermal analysis of temperature distribution resulting from waste heat discharges. Three other important heat transport processes are heat storage within the water column, heat advected with currents, and heat dispersed by turbulent transport processes. Surface heat exchange is the summation of seven separate interfacial heat transfer processes. These processes are: incoming short-wave radiation, long-wave atmospheric radiation, reflection of short- and long-wave radiation, back radiation from the water surface, evaporation, and conduction. The seven surface heat exchange mechanisms are illustrated in Figure 1. Their algebraic sum gives the net rate of surface heat exchange defined as:

$$H_n = (H_s + H_a - H_{sr} - H_{ar}) - (H_{br} + H_e + H_c) \tag{1}$$

The net rate of surface heat exchange is divided into the two groups of terms of incoming radiation minus their reflective components, and three water surface temperature dependent heat loss terms. Evaluation of surface heat exchange for use in temperature predictions depends on the details of the three heat loss processes.

Each of the three water surface temperature-dependent heat loss processes can be evaluated in excruciating detail. Evaluation of evaporation is presented in a paper by Wilfried Brutsaert (1975). His paper clearly illustrates the present state of detailed understanding of the evaporative process as well as the difficulties of evaluation. Its nighttime counterpart, condensation at the water surface, remains unevaluated as it influences the net rate of surface heat exchange. Each of the remaining components

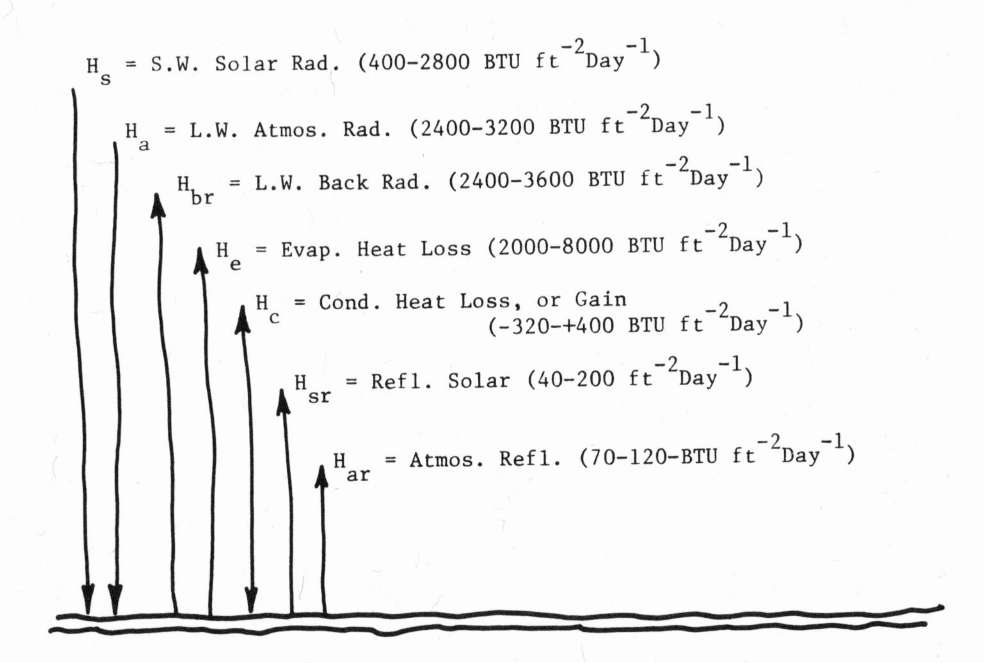

NET RATE AT WHICH HEAT CROSSES WATER SURFACE

$$H_n = (H_s + H_a - H_{sr} - H_{ar}) - (H_{br} \pm H_c + H_e) \text{ BTU ft}^{-2}\text{Day}^{-1}$$

H_R — Absorbed radiation independent of temp.

Temp. dependent terms

$$H_{br} \sim (T_s + 460)^4$$

$$H_c \sim (T_s - T_a)$$

$$H_e \sim W(e_s - e_a)$$

Figure 1. Mechanisms of heat transfer across a water surface.

of the net rate of surface heat exchange can be examined in equally voluminous detail and are as equally uncertain as evaporation in their evaluation.

For example, reflection of incoming radiation is considered to be a small fraction of the incoming radiation, on the order of 10%. It is thought to be a function of the roughness of the water surface, the angle of the sun, and the type of cloud cover. There is also some uncertainty in the determination of long-wave atmospheric radiation.

Back radiation from the water surface is considered to be a fourth-power radiation process following the Stephan-Boltzman equation with constant emissivity. However, there is the outstanding question of exactly what constitutes the water surface temperature at which the back radiation takes place. It is thought to be the temperature of a millimeter thick surface layer yet for temperature prediction purposes it is assumed to be at the temperature of a finite thickness of the water column.

Conduction of sensible heat from and to the water surface is modified by the formation of fog and condensate in the air immediately above the water surface. Its evaluation is as complex as the evaluation of evaporation and is not sufficiently understood as a separate heat transfer mechanism.

The dilemma that arises in the hydrothermal analysis problem is one of determining the role of surface heat exchange in the temperature prediction problem without being limited by the lack of knowledge of the individual surface heat exchange processes. Since the hydrothermal analysis problem is one of temperature prediction, the closer the evaluation of the net rate of surface heat exchange can be brought to the basic parameter being predicted (water temperatures) the easier it might be to develop the complex prediction equations required for the study of detailed temperature distributions.

Since certain of the terms in Equation 1 are surface temperature dependent, and others are measurable or computable input variables, the most direct route is to define an equilibrium temperature, E, as the temperature at which the net rate of surface heat exchange is zero. The equilibrium temperature becomes the fictitious water surface temperature at which the incoming radiation heat rates are just balanced by the outgoing water surface temperature dependent processes. Linearization of Equation 1 along with the definition of equilibrium temperature allows expressing the net rate of surface heat exchange, H_n, simply as:

$$H_n = -K(T_s - E) \qquad (2)$$

where K is the coefficient of surface heat exchange, and T_s is the water surface temperature. Seven separate heat exchange processes are summarized in these two parameters which are the coefficient of surface heat exchange and the equilibrium temperature. The linearization used in obtaining Equation 2 has been examined in detail by Brady, et al. (1968), and Edinger et al. (1974).

The definition of the coefficient of surface heat exchange can be shown to be the first term of a Taylor series expansion by considering Equation 2 as:

$$H_n = -\frac{dH_n}{dT_s}(T_s - E) \tag{3}$$

where the derivative of H_n with respect to surface temperature is evaluated from Equation 1 to give K, the coefficient of surface heat exchange. All the approximations of the individual surface heat exchange terms enter into the evaluation of the coefficient of surface heat exchange and the equilibrium temperature. Equations 2 and 3 are defined from Equation 1; they have the same difficulties of evaluation as the individual terms in Equation 1; but they provide a simpler algebraic device for inclusion of the net rate of surface heat exchange in complex temperature prediction analyses.

The hydrothermal analysis problem allows another simplifying step that makes the linearization of the net rate of surface heat exchange more precise than implied by Equations 2 and 3. For purposes of evaluating the increment or excess temperature due to man-made heat sources, the excess temperature can be defined as a perturbation of the ambient water temperature producing a new net rate of surface heat exchange, H_n', at an elevated temperature of $T_s + \Theta$, giving:

$$H_n' = -K(T_s + \Theta - E) \tag{4}$$

The increment of surface heat exchange due to a heat source becomes from Equations 2 and 4:

$$H_n' - H_n = -K\,\Theta \tag{5}$$

and allows examining only the transport of excess heat. The increment of surface heat exchange due to a heat source can be evaluated without regard to the knowledge of the equilibrium temperature and with only a vague understanding of the overall heat budget of a water body.

Direct evaluations of the net rate of surface heat exchange under conditions of heat loadings has been carried out by Brady, et al. (1968). Empirical evaluation of Equation 1 has led to the following relatively simple equation for equilibrium temperature:

$$E = \frac{H_s}{K} + T_d \tag{6}$$

in which H_s is the rate of incoming short-wave radiation and T_d is the atmospheric dewpoint temperature. For diurnal cycles of heat transfer, it has been found that the first, or short-wave radiation term dominates. On a diurnal basis, the equilibrium temperature can have an amplitude as great as 25°C due to short-wave solar radiation, Edinger et al. (1968). On an annual cyclic basis, however, it is found that the dewpoint variation dominates and that the equilibrium temperature has an annual amplitude slightly larger than the annual variation in dewpoint temperature.

Incorporation of surface heat exchange in hydrothermal analysis changes the scientific question of how to better understand each term in Equation 1 to a practical one of asking how precisely the net rate of surface heat exchange needs to be known to make adequate excess temperature predictions associated with large sources of waste heat. Rather than consider the accuracy of the net rate of surface heat exchange relative to the individual mechanisms of radiation, conduction and evaporation, the question becomes one of examining the magnitude of surface heat exchange relative to the heat transport processes of advection and dispersion, as they occur in complex excess temperature prediction relationships. The need for the approximations to the net rate of surface heat exchange become clearer as the details of the hydrothermal problem are examined.

It is now recognized that there are three scales associated with the mixing processes that enter into the development of heat transport relationships. There is the local or initial mixing scale that is primarily dependent on the specific way the waste heat is introduced into the receiving water body. This scale spreads heat over tens of acres of surface area. There is the intermediate scale describing the advection and dispersion of heat by ambient currents and turbulence within the water body and covering hundreds to thousands of acres of surface area. Third, there is the large scale concerned with the total heat budget of a water body and its perturbation by the waste heat sources, and can cover miles of length of a river or estuary or many square miles of a lake or coastal area. In addition to the three areal scales there are three spatial dimensions over which the heat transport processes can be approximated for prediction purposes. The three scales and three associated spatial dimensions are outlined in Table 1 along with the generic name of types of models that have been developed to apply at each combination of scale and dimension. Any temperature prediction problem reduces to a choice of one of the models from each of the scales. The permutations in

Table 1 give over 1,000 possible combinations of models. Table 1 can be further multiplied by available solution techniques ranging from simple rules of thumb to complex computer simulations. It becomes readily evident why the hydrothermal analysis problem quickly degenerates to a site by site study of the particular mechanisms operating at a particular site.

Experience has taught that surface heat exchange is relatively unimportant at the initial or local mixing scale, can have some importance at the intermediate scale, and is dominant at the large scale. This can be shown to be the case by examining simplified analyses that apply to each scale when the temperature distributions resulting from the advection and dispersion processes are described in terms of surface area. Surface area descriptions of the intermediate mixing processes were first developed in Edinger and Polk (1971), and have been utilized by Pritchard and Carter (1971) in making first order approximations to initial mixing processes, and have been extended through all three scales of analysis by Edinger et al. (1974).

Initial mixing can be approximated by a momentum balance of a linearly expanding jet for which the dilution of this jet centerline excess temperature is given as:

$$\frac{\Theta}{\Theta_0} = \left(\frac{z}{z_0}\right)^{-\frac{1}{2}} \qquad (7)$$

where Θ_0 is the initial excess temperature due to the heat source, z_0 is the virtual source distance and is related to discharge geometry, and Θ is the temperature excess at a distance z along the jet axis. Initial mixing for a simple plane jet results in a ½ power law for temperature dilution. The linear expansion law of the jet is given by:

$$\frac{b}{b_0} = \frac{z}{z_0} \qquad (8)$$

where b_0 is the initial jet width and is a function of discharge design. Equation 8 results from the hydrodynamic similarity relationships, and Equation 7 results from applying the linear expansion law to the combined momentum and heat balance of the jet. With no surface cooling, the surface area within a given temperature contour during initial mixing is given by:

$$A = \frac{z_0 b_0}{2} \left[(\Theta_0/\Theta)^4 - 1 \right] \qquad (9)$$

The surface area for the plane jet increases with the fourth-power of the excess temperature ratio. A typical 1,000 MW plant would discharge 1,000 cfs at a 15°F temperature rise. If the

TABLE 1

CLASSIFICATION OF TYPES OF MODELS IMPORTANT IN HYDROTHERMAL ANALYSIS REGARDLESS OF SOLUTION TECHNIQUES

Scale	One Dimensional Cases	Two Dimensional Cases	Three Dimensional Cases
1. Localized discharge design		1. Plane Jet 2. Plane Jet with friction 3. Deflected Jet 4. Hydraulically stratified discharges 5. Effects of oscillating currents	1. Axisymetric Jet 2. Buoyant Jet 3. Deflected Jet 4. Submerged buoyant jet, single port 5. Submerged buoyant jet, multiport 6. Effects of oscillating currents
2. Intermediate mixing cases	1. Simple river dilution	1. Longitudinal advection with lateral dispersion, no shoreline	1. Longitudinal advection, lateral and vertical dispersion, no shoreline
	2. Sectionally homogeneous estuary dilution	2. With shoreline	2. With shoreline
	3. Sectionally homogeneous and one dimensional dispersion relationships, including oscillating currents	3. Bounded two shorelines	3. Other combinations of boundary conditions
		4. Longitudinal advection with vertical dispersion, bounded	4. Above with computed velocity fields including oscillating currents
		5. Other combinations of boundary conditions 6. Above with computed velocity fields including oscillating currents	
3. Large scale and time varying distributions	1. Advection with heat exchange, cooling lakes, simple river	1. Longitudinal and lateral advection and dispersion with two dimensional hydraulics	1. Generalized three dimensional hydro-dynamic problem, including buoyancy and convection
	2. Sectionally homogeneous estuary	2. Two dimensional planar flows with stratified interface	2. Merger with other scales
	3. Sectionally homogeneous analysis with time varying hydraulics	3. Two layered approximations to estuarine circulation, and cooling lake convection	
	4. Planar homogeneous lakes & reservoirs	4. Hydrodynamics of longitudinal and vertical motions with buoyant convection	
	5. Stratified flow hydraulics		

discharge orifice had a width of 20 feet and the virtual source distance, z_0, is taken as $6b_0$, the area within the jet region to a dilution of 5^oF would be 96,000 ft^2 or about 2.5 acres. If surface heat exchange were taking place over this area (with a typical coefficient of surface heat exchange of 125 Btu $Ft^{-2}Day^{-1}$ $^oF^{-1}$ and for a temperature rise of 15^oF) the total heat lost would be 0.18×10^9 Btu/Day; but the total heat rejection of the plant is 79×10^9 Btu/Day. Even if initial mixing took place over ten times the calculated surface area, the total heat lost to the atmosphere within the initial mixing region would be less than 3% of the heat rejected.

Relationships for the intermediate region can be expressed in terms of surface area relationships that can be derived from the basic advection and dispersion heat balance. Surface heat exchange can be included in the vertically mixed case. The theoretical result of such an analysis is shown in Figure 2 which gives the excess temperature ratio as a function of the ratio of surface area to a scaling area, and as a function of a surface heat exchange parameter, β. The scaling area, A_n, is related to the plant pumping rate, the lateral dispersion coefficient, the depth of the water column, and the ambient current velocity. The surface heat exchange parameter is given by:

$$\beta = \frac{\sqrt{\pi}}{4} \frac{K A_n}{\rho c_p Q_s} \qquad (10)$$

where ρ and c_p are the specific weight and specific heat of water respectively, and Q_s is the condenser flow rate. A typical value of the scaling area for a discharge plume in a small estuary is $1.5 \times 10^6 ft^2$. For the hypothetical plant this gives a value of β = 0.015 for the curve of Figure 2. Using this value of β in Figure 2, it is found that surface cooling has little influence on A/A_n until Θ/Θ_0 is less than about 0.2. At Θ/Θ_0 = 0.2 on Figure 2, it is found that A/A_n is approximately 12 for β = 0.015 and approximately 25 for $\beta = 0$ (no consideration of surface cooling). Integrating the area temperature curve for a coefficient of surface heat exchange of K = 125 Btu $Ft^{-2}Day^{-1} {}^oF^{-1}$ gives a total heat loss throughout the intermediate region of 16.7×10^9 Btu Day^{-1} which is 21% of the total heat rejection. Inclusion of surface cooling at the intermediate scale can reduce the area required to reach low values of excess temperature by a factor of two. Surface cooling, however, has little influence on the surface area contained within the higher values of Θ/Θ_0 near the discharge.

The principle that surface cooling is important only over large surface areas and at lower values of excess temperature is further illustrated by considering the cooling process alone. When surface heat exchange is balanced against simple advection as in a cooling pond, the resulting expression is the exponential temperature

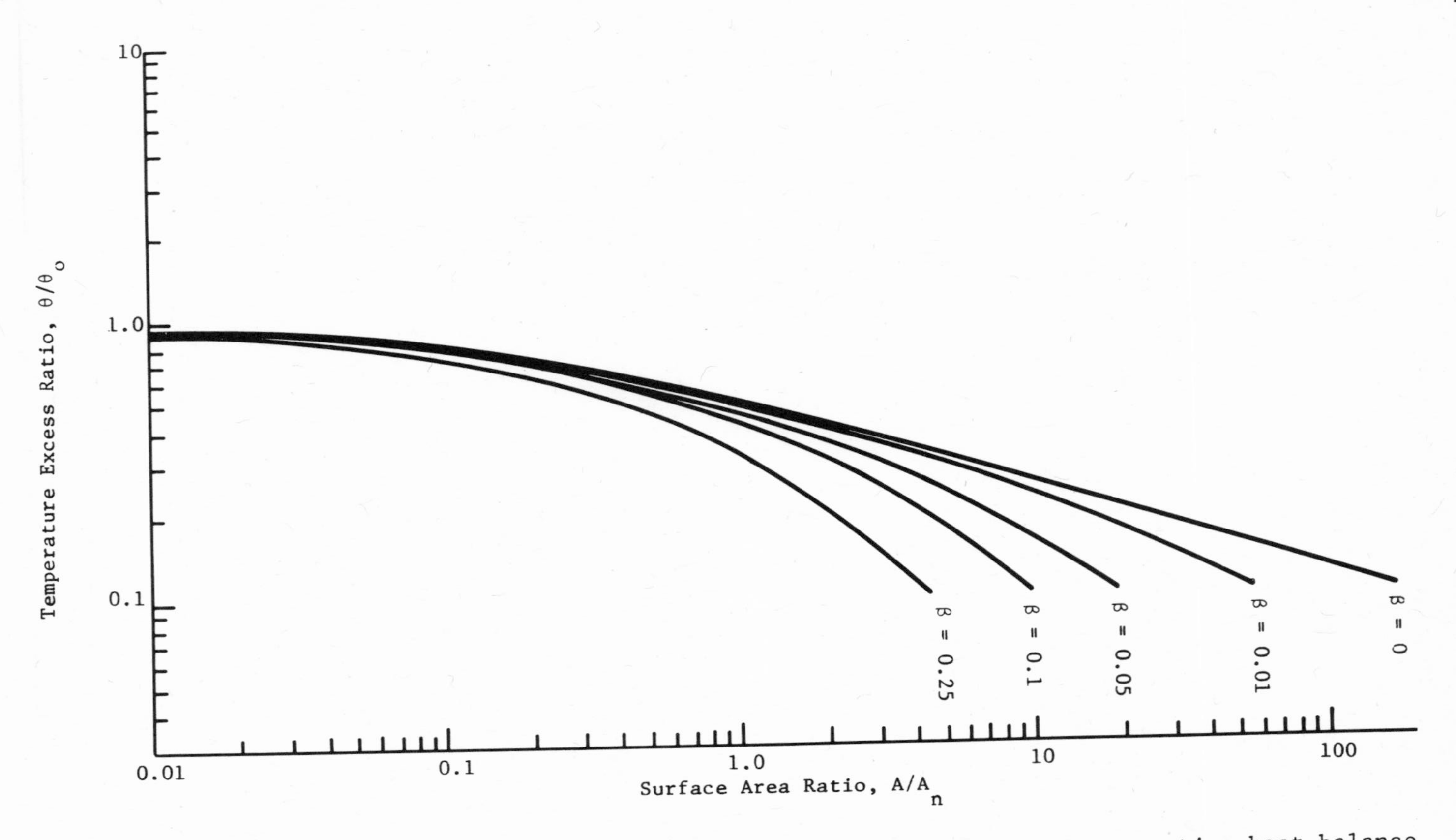

Fig. 2. Excess temperature versus areal spread for two-dimensional non-conservative heat balance.

decay:

$$\Theta = \Theta_0 \exp\left[-KA/\rho c_p Q_s\right] \tag{11}$$

which gives a required surface area of $68 \times 10^6 ft^2$ to reach a Θ/Θ_0 of 0.20. This area is at least 10 times the area within the intermediate mixing region and almost 1,000 times the surface area within the initial mixing or jet region.

The above order of magnitude estimates of area required by each of the scales of processes and the importance of surface cooling are based on the assumption that each region is independent of the other region. The real problem is more complex, as illustrated by examining Table 1, since the end of the initial mixing region establishes the value of Q and Θ to be used in the intermediate region computations, and the end of the intermediate region sets the value of Θ_0 for the large scale process, as for example Equation 11.

Eventually, all of the excess heat due to a cooling water discharge is dissipated to the atmosphere. Most of the surface heat dissipation takes place at low values of temperature excess, on the order of 2^oF to 3^oF for the above example case, at a net rate of approximately 300 Btu $Ft^{-2}Day^{-1}$. Compared to the orders of magnitude of the natural exchange processes given in Figure 1 this net rate is about 10% of the average daily energy input due to solar radiation or long-wave atmospheric radiation. It is also about 10% of the average daily heat loss due to evaporation, and is of the same order as the heat conduction and reflected energy terms. It is very difficult to detect the increment of surface heat exchange due to a cooling water discharge into a large open water body where advection and dispersion processes dominate over a wide expanse of the plume.

Accurate evaluation of the increment of surface heat exchange due to a cooling water discharge has been achieved in closed water bodies such as cooling ponds. The boundaries confine the heat and recirculation through the plant raises the lake excess temperature to a sufficient level such that the increment or excess of surface heat exchange is a significant part of the total surface heat exchange. The heat balance for a completely mixed pond is:

$$H_p = K\Theta A \tag{12}$$

where H_p is the heat rejection rate of the power plant, Θ is the temperature excess of the pond, and A is the pond surface area. For the heat rejection rate of 79×10^9Btu Day^{-1} and a coefficient of surface heat exchange of 125 Btu $Ft^{-2}Day^{-1}{}^oF^{-1}$, Equation 12 gives a required surface area of $0.26 \times 10^8 Ft^2$ or 2900 acres at a 5^oF excess temperature, and $0.63 \times 10^8 Ft^2$ or 1450 acres at a 10^oF

excess temperature. It is the latter figure that results in the rule of thumb that about 1.5 acres of cooling pond area is needed for each MW of plant capacity. It would be more correct to say that 1.5 acres is required to cool 1 MW at a $10^{\circ}F$ increment of excess temperature. This varies slightly as the exchange coefficient varies. The rule obviously does not apply to open water bodies where dilution is available. It is not possible in the presence of mixing and dilution processes to build up head loadings on the order of 2/3 MW per acre. A reasonable order of magnitude for open water bodies is more like 1 MW per 10 acres with most of the dissipation taking place over large surface areas at small increments of excess temperature.

In open water bodies such as lakes and coastal waters, the excess heat due to a cooling water discharge is dissipated at approximately the same rate as the natural rate of net surface heat exchange. This rate is of the order of the error in establishing the overall heat budget of the water body under natural conditions. A major factor that inhibits the detection of this small increment in the natural surface heat exchange processes is heat storage within the water column on a diurnal basis. Many days of plume data must be averaged to detect an increment of $2^{\circ}F$ or $3^{\circ}F$ resulting from a cooling water discharge, and even this determination is uncertain in the presence of daily varying currents due to winds or tides.

Semiclosed water bodies such as rivers or estuaries present an upper limit to initial and intermediate mixing of a thermal discharge as determined by the completely mixed excess temperature. For a balance between longitudinal advection, dispersion and surface cooling, the fully mixed excess temperature can be expressed as:

$$\Theta_m = \frac{H_p}{\rho c_p \left[Q^2 + \frac{4KWSD_x}{\rho c_p} \right]^{\frac{1}{2}} + K\,A_m} \tag{13}$$

where W is the width of the water body, S is the water body cross-sectional area, D_x is the longitudinal dispersion coefficient, and A_m is the surface area of the water body over which complete mixing to Θ_m takes place by initial and intermediate processes. The flow rate Q is the net fresh water flow in the case of an estuary. Equation 13 expresses the completely mixed temperature for the whole spectrum of cases of a cooling pond, a river flow situation, an estuary where river flow and longitudinal dispersion contribute about equally to the mixing, and a coastal water situation where longitudinal dispersion dominates the complete mixing process.

The different cases for which Equation 13 applies are determined by the relative magnitudes of Q, $KA_m/\rho c_p$ and SD_x. If Q is much larger than SD_x and $KA_m/\rho c_p$, then Equation 13 reduces to the simple river dilution case. If SD_x is much larger than Q and $KA_m/\rho c_p$, then longitudinal dispersion dominates as would be the case in the ocean reaches of an estuary or an enclosed tidal embayment with little fresh water inflow. If $KA_m/\rho c_p$ dominates over Q and SD_x, then Equation 13 reduces to Equation 12 with surface heat dissipation being dominant from the region over which mixing takes place. The six possible cases given by Equation 13 are summarized in Table 2 for each possible combination of processes. This comparison of the combined terms give some hope of predetermining the dominant processes for different water bodies based upon their geometry and dispersion characteristics. It is clear from Equation 13 that surface heat exchange becomes the limiting heat dissipation process in two out of the three terms.

The overlap and intersection of the different scales of heat advection, dispersion and dissipation is illustrated for an estuarine discharge in Figure 3. The plant discharge is a tidal canal with little momentum at the edge of the water body, hence the heat begins to disperse by the intermediate processes. The intermediate region in Figure 3 is bounded by the flood and ebb conditions from Θ/Θ_0 of unity to Θ/Θ_0 of approximately 0.20. Excess temperatures in this region were large enough to determine the plume areas from observed temperatures. The flood and ebb tide bounding curves were determined from the type of relationships illustrated in Figure 2.

Table 2

COMBINATIONS OF DOMINANT CASES FROM THE ADVECTION, DISPERSION, AND SURFACE COOLING FULLY MIXED TEMPERATURE HEAT BALANCE

	c_pQ	$4KWSD_x$	KA_m
c_pQ	River dilution	Tidal estuary advection and dispersion	Plug flow cooling pond
$4KWSD_x$		Embayment tidal mixing alone	Embayment mixing with limited surface area
KA_m			Completely mixed cooling pond

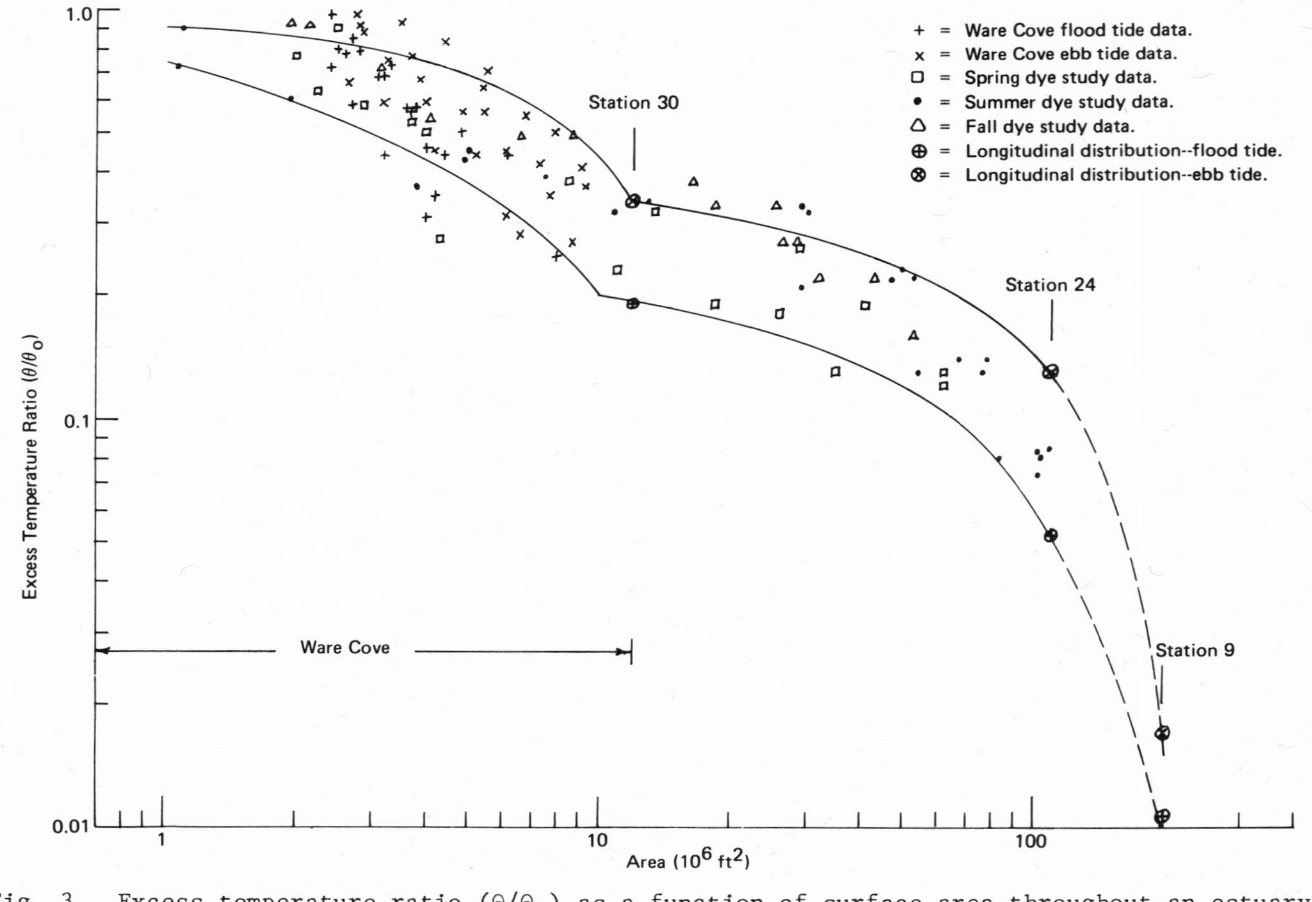

Fig. 3. Excess temperature ratio (Θ/Θ_0) as a function of surface area throughout an estuary.

The large scale processes in Figure 3 begin for Θ/Θ_0 of less than 0.2. The bounding ebb and flood tide curves were derived from a two-layered estuarine circulation analysis similar to that given in Edinger, et al. (1974). Figure 3 shows a sharp transition between the intermediate scale processes and the large scale processes as heat is spread across the water body to be included in the general longitudinal circulation. The large scale processes begin with little change in Θ/Θ_0 over a large region of surface area followed by a gradual decrease as surface heat dissipation becomes dominant at the larger areas. The data points between the bounding curves of the large scale processes were determined from dye tracer studies which were required to detect the excess heat remaining at low values of excess temperature.

Surface heat exchange in hydrothermal analysis is not only a large surface area process, it is also a long term process. This can be seen by examining the time required for the excess temperature to build up in a simple cooling pond. The relationship between the increase in heat storage and surface heat exchange for a pond of volume V and surface area A is given by:

$$\rho c_p V \frac{d\Theta}{dt} + K\, A\, \Theta = H_p \tag{14}$$

In the limit, Equation 14 approaches the fully mixed temperature given by Equation 12. Signifying this latter value by Θ_m, then for a constant rate of heat loading, the time dependent warming-up temperature from Equation 14 is given by:

$$\Theta(t) = \Theta_m \left[1 - \exp(-Kt/\rho c_p d) \right] \tag{15}$$

where d is the average pond depth or V/A. From Equation 15, it is found that for a 10 foot average pond depth, it takes about 11 days to reach 90% of Θ_m. An average- sized cooling pond has a residence time of about 5 days. Two recirculations of the heat are therefore necessary to build up the surface heat loading to the level at which surface heat dissipation balances the rate of heat rejection by the cooling water. Very little heat build-up takes place within the first day which is analogous to the time of travel through the intermediate mixing region of an open water body.

Other comparisons of the magnitude and role of surface heat exchange in hydrothermal analysis can be made; but they all lead to the same conclusion, which is that surface heat exchange is a large area and long term process in comparison to mixing, advection, dispersion for a relatively large waste heat source.

It is possible to make these order of magnitude comparisons in a relatively straight forward manner because of the linear approximation to the term by term surface heat exchange processes expressed in Equation 1. It is also evident that the linear approximation is sufficiently accurate for comparison of the smaller area and shorter time processes of mixing, advection, and dispersion. It would have been very difficult to arrive at these illustrations and conclusions if the term by term description of surface heat exchange were employed.

REFERENCES

Brutsaert, W., "Evaporation from Water Surfaces," in Interfacial Transfer Processes in Water Resources, Water Rescurces and Environmental Engineering Research Report No. 75-1, Dept. of Civil Engineering, University of Buffalo, (1975).

Edinger, J. E., D. W. Duttwieler, and J. C. Geyer, "The Response of Water Temperatures to Meteorological Conditions," Water Resources Research, 4, No. 5, (1968).

Edinger J. E. and E. M. Polk, "Intermediate Mixing of Thermal Discharges into a Uniform Current," Water, Air and Soil Pollution, 1, No. 1, (1971).

Edinger, J. E., D. K. Brady, and J. C. Geyer, "Heat Exchange and Transport in the Environment," Edison Electric Institute, Cooling Water Discharges Project Report No. 14, New York (1974).

Brady, D. K., W. L. Graves, and J. C. Geyer, "Surface Heat Exchange at Power Plant Cooling Lakes," Edison Electric Institute Report No. 69-901, New York (1969).

Pritchard, D. W. and H. H. Carter, "Design and Siting Criteria for Once-Through Cooling Systems Based on a First Order Thermal Plume Model," Report No. 75, Chesapeake Bay Institute, Johns Hopkins University, Baltimore, Md., (1972).

NOTATION

A = pond surface area
A_m = surface area of the water body over which complete mixing to Θ_m takes place
A_n = scaling area
β = surface heat exchange parameters
b = jet width
b_0 = initial jet width
c_p = specific heat of water
d = average pond depth
D_x = longitudinal dispersion coefficient
E = equilibrium temperature
H_a = long-wave atmospheric radiation
H_{ar} = atmospheric reflection
H_{br} = long-wave back radiation
H_c = conduction heat loss
H_e = evaporation heat loss
H_n = net surface heat exchange
H_n' = net rate of surface heat exchange due to excess temperature
H_p = heat rejection rate of the power plant
H_R = absorbed radiation
H_{sr} = reflective solar radiation
K = coefficient of surface heat exchange
Q_s = condenser cooling water flow rate
S = water body cross-sectional area
T_d = atmospheric dewpoint temperature
T_s = water surface temperature
V = volume of water body
W = width of water body
z = coordinate direction
z_0 = virtual source distance
ρ = specific weight of water
Θ = temperature excess of the pond
Θ_0 = initial excess temperature due to heat source
Θ_m = mixed water body temperature

DISPERSION OF POLLUTANTS IN OPEN-CHANNEL LAMINAR FLOW

H. P. Hsieh, Gi Yong Lee, and William N. Gill

Faculty of Engineering and Applied Sciences, State

University of New York at Buffalo

ABSTRACT

Generalized dispersion theory is used to study the unsteady dispersion of pollutants in open-channel flow. It is shown that the dispersion equation has time dependent coefficients and that it applies only to initial value problems such as the concentration distribution created by an instantaneous source.

A superposition integral is used to study the distribution of BOD from a continuous source. It is shown that the usual constant coefficient dispersion model can lead to significant errors when applied to continuous source problems, especially when one is interested in concentration distributions near the source.

INTRODUCTION

With the growing demands on water resources, all phases of environmental control, particularly water pollution control, have gained tremendous importance in recent years. To control pollution, the disposal of industrial, agricultural and domestic wastes into waterways should be regulated. This requires knowledge of how the released pollutants are dispersed in the system. It is intended in this work to investigate dispersion in open-channel flow using a new method of analysis which, it is hoped, will contribute to better understanding the phenomena associated with the dispersion of material pollutants.

Longitudinal dispersion is the action by which a flowing stream spreads out and dilutes a mass of miscible pollutants. The velocity variation over the cross-section of the flow distorts and enlarges the region the pollutant occupies in the direction of flow and hence enhances the longitudinal dispersion. Usually, transverse diffusion tends to compress the mixing zone of the pollutant and thereby to inhibit the longitudinal dispersion. Longitudinal diffusion is another mechanism which contributes to longitudinal dispersion, but in many cases this contribution is minor.

The first important study of dispersion was done by G. I. Taylor (1953, 1954) for laminar and turbulent flows in a straight tube. He presented an approximate mathematical description of the unsteady convective diffusion of solute in flowing streams in terms of the following one-dimensional diffusion model:

$$\frac{\partial C_m}{\partial t} + u_{avg} \frac{\partial C_m}{\partial x} = k_2 \frac{\partial^2 C_m}{\partial x^2} \tag{1}$$

where C_m is the area-average concentration of solute given by

$$C_m = \frac{\int_0^A C dA}{\int_0^A dA} ,$$

t is the time after the solute is added to the flow, x is the longitudinal distance from the point of origin at which the solute was introduced, u_{avg} is the average velocity of the flow, k_2 is the dispersion coefficient and A represents the cross-sectional area of the flow. Unlike the molecular diffusion coefficient, the dispersion coefficient is not a unique physical property of the fluid but depends on the fluid dynamics and the physical parameters of the system. Taylor (1953) determined the asymptotic value of k_2, valid only for large values of time, in laminar fully developed tube flow to be

$$k_2 = \frac{u_{avg}{}^2 R^2}{48D} \tag{1a}$$

where R is the tube radius and D is the molecular diffusivity of the solute in the fluid. For fully developed turbulent flow, Taylor (1954) obtained

$$k_2 = 10.1 R u_* \tag{1b}$$

where $u_* \equiv \sqrt{\tau_w/\rho}$; τ_w is the shear stress at the wall, and ρ is the fluid density.

Aris (1956) used the method of moments to solve the unsteady convective diffusion equation approximately. He generalized Taylor's result to include longitudinal molecular diffusion and obtained

$$k_2 = \frac{u_{avg}{}^2 R^2}{48D} + D \quad . \tag{1c}$$

Elder (1959) applied Taylor's reasoning to turbulent flow in an infinitely wide open channel and found the dispersion coefficient, using a logarithmic velocity profile, to be

$$k_2 = 5.93 y_m u_* \tag{1d}$$

where y_m is the depth of channel flow. Sayre (1967) solved the two-dimensional equation for open-channel turbulent flow with the logarithmic velocity profile. He employed the method of moments and verified Elder's result for large times. Fischer (1966) conducted experiments in two-dimensional laboratory flumes and found that the observed dispersion coefficients showed a remarkable uniformity but were about 40% higher than Elder's calculation indicated. Fischer attributed these discrepancies to the value of the von Karman constant used (0.41 by Elder vs. 0.36 by Fischer), the deviations from the assumed logarithmic velocity profile and mixing coefficient, and that the flow was not strictly two-dimensional.

Ananthakrishnan, Gill, and Barduhn (1965) studied dispersion in laminar tube flow and mapped the regions in which convection dominates and when Taylor's theory, with or without Aris' modification, applies. They carried out extensive numerical calculations to determine the regions in which various mechanisms control the process.

Fischer (1966) determined from both his analysis and experiments that there were three distinctive time periods in the dispersion of pollutants. These are: (1) the convection period when Taylor's theory is definitely not applicable; (2) the transition period when the variance grows linearly with time but equation (1) is not applicable; and (3) the Taylor diffusion period when equation (1) applies. Sayre (1967) separated the dispersion process into an initial period and an equilibrium period. His equilibrium period corresponds with values of time large enough for the Taylor diffusion theory, equation (1), to apply and it occurs after a dimensionless dispersion time of

$$\tau' = \frac{\varepsilon_{avg} t}{y_m^2} \gtrsim 0.5$$

where ε_{avg} is the cross-sectional average of the vertical eddy diffusivity.

Thus, several investigators proved that Taylor's one-dimensional diffusion theory applies only after a certain time has elapsed, a conclusion that Taylor also reached in his original work. However, they were not able to present any analytical predictive methods to account for the period before the Taylor diffusion theory becomes applicable. Furthermore, Taylor's dispersion model applies rigorously only to physical situations in which a finite amount of material or energy is introduced into the system at $t = 0$. It does not apply directly when material or energy are introduced continuously as is done in many practical applications. Many investigators (Fischer, 1966; Glover, 1964; Godfrey and Frederick, 1963; Thomas, 1958) have observed that the experiments in three-dimensional laboratory flumes and field tests in natural streams yielded dispersion coefficients much larger (up to two orders of magnitude) than Elder's. Fischer proposed a model to predict the larger dispersion coefficient in natural streams using Taylor's concept. He reasoned that the governing mechanism for longitudinal dispersion in natural streams was the lateral instead of the vertical variations of velocity. He found good agreement between his model and field tests.

Unsteady convective diffusion in open-channel flows has been analyzed by a few investigators from points of view other than the dispersion model. Hays (1966) proposed a dead-zone model, similar to that suggested by Turner (1958), to account for the long-tailing effect of the pollutant cloud as demonstrated in Figure 1. Hays divided a typical cross section of a stream or river into two distinctive zones: a main stream and a dead zone along the bed and banks and set up a one-dimensional plug flow model,

$$\frac{\partial C}{\partial t} = D_a \frac{\partial^2 C}{\partial x^2} - u_{avg} \frac{\partial C}{\partial x} - Ka(C_d - C)$$

for the main stream and a first order reaction model,

$$\frac{\partial C_d}{\partial t} = Kd(C - C_d)$$

for the dead zone which is assumed to be perfectly mixed, and

D_a = dispersion coefficient in the main stream

K = mass transfer coefficient

a = ratio of contact area to main stream volume

d = ratio of contact area to dead zone volume

C_d = concentration in the dead zone.

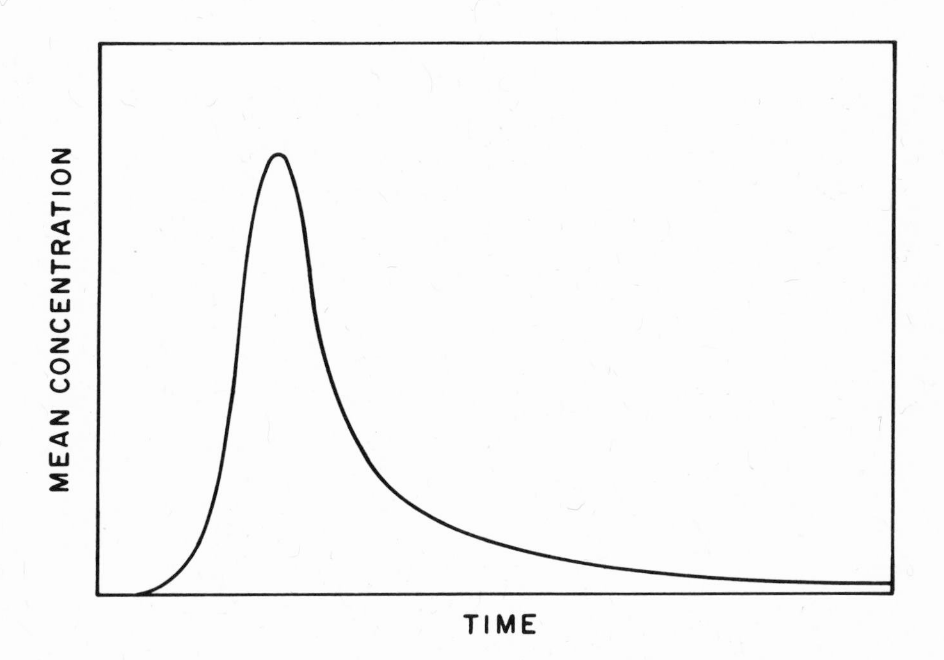

Figure 1. Long-tailing effect of pollutant cloud.

Hays solved the coupled differential equations using the Laplace transform for a pulse input and claimed the superiority of this model over Taylor's model in predicting the long-tailing effect. Thackston and Krenkel (1967) illustrated the applicability of this dead-zone model through laboratory and field experiments. However, K and D_a must be regarded as empirical constants.

Yotsukura and Fiering (1964) solved the two-dimensional unsteady convective diffusion equation,

$$\frac{\partial C}{\partial t} + u(y) \frac{\partial C}{\partial x} = \frac{\partial}{\partial y} \varepsilon_y(y) \frac{\partial C}{\partial y}$$

numerically for the same flow conditions studied by Thomas (1958) and Elder (1959) using the logarithmic velocity distribution. But Fischer (1965) found that their asymptotic dispersion coefficient was not in agreement with the work of Thomas and Elder. Patterson and Gloyna (1965) proposed an empirical expression for the dispersion coefficient in the form of

$$k_2 = P_1 \left[\frac{u_{avg}}{\ln A_r}\right]^{P_2}$$

where

P_1 = 0.258 for laboratory flume and 229 for natural streams

P_2 = 0.830 for laboratory flume and 0.269 for natural streams

A_r = aspect ratio, width divided by depth of flow

based on the available experimental data, but the fit was not satisfactory and the relation is not dimensionally homogeneous (Fischer, 1966). Note that the dispersion coefficient in natural streams could be roughly 1,000 times as big as the one in laboratory flumes.

Another important problem related to material pollution and dispersion is the prediction of the biochemical oxygen demand (BOD) distribution in a flowing stream which results from the discharge of a continuous source. It is important to predict the BOD distribution for the following reasons: (1) BOD is present in practically all forms of waste which are released into waterways and (2) the BOD distribution is closely related to the oxygen distribution which, in turn, affects the reaeration rate.

It has been assumed by previous investigators, primarily for simplicty, that BOD is satisfied according to a first order, homogeneous reaction independent of the concentration of oxygen.

Table 1 shows a summary of the proposed models to predict BOD concentration in flowing streams. Biguria et al. (1969) and Ahlert (1971) separated the BOD reaction into two stages: carbonaceous and nitrogenous.

Gill (1967) proposed a method of solving the unsteady convective diffusion equation which employed an infinite series expansion involving derivatives of the area-mean concentration. This approach will be explained in detail later.

Gill and Sankarasubramanian (1970) derived a generalized dispersion model based on the series solution developed earlier by Gill (1967). This generalized dispersion model describes the unsteady tracer concentration for smaller values of time than Taylors model and an essential feature of it is that its coefficients depend on time in a predictable way. Gill and Sankarasubramanian (1971) extended their work to account for non-uniformity in the distribution of the source. They further generalized the technique to handle continuous sources by using a superposition integral (Gill and Sankarasubramanian, 1972a; Sankarasubramanian and Gill, 1972b). In this series of articles, laminar flow in a tube was treated as a specific example. Subramanian et al. (1974) extended the generalized dispersion model to include first order homogeneous chemical reaction as in the case in BOD studies. Sankarasubramanian and Gill (1973,1974) and Gill (1975) used the generalized dispersion theory to study heterogeneous first order reactions in laminar flows. The same approach has been used by Nunge (1974a,1974b) and Nunge and Subramanian (1977) to study atmospheric dispersion.

The purpose of this article is to extend the work of Gill and Sankarasubramanian (1967,1970,1971,1972a,1972b) to two-dimensional rectangular open-channel flows. The thesis of Hsieh (1971) forms part of the basis of this report in which only laminar flow will be treated. However, as mentioned earlier, the work can be extended readily to turbulent systems but the computations may be formidable depending on the eddy diffusivity used.

The unsteady convective diffusion problem with a given initial concentration distribution will be treated and the non-uniformly distributed pulse input (instantaneous source) of pollutant to the system will be taken as a specific case. It will be shown that the convective coefficient and the dispersion coefficient in the dispersion model are strong functions of time in the early stage of the dispersion process and the dependence of the coefficients on time will be derived explicitly.

The non-uniformly distributed continuous source problem can be solved by using the superposition integral. It will be shown that, for a given discharge rate of pollutant into the system, one may achieve greater dilution of the pollutant by placing the source

TABLE 1. A SUMMARY OF MODELS FOR BOD PREDICTION

Investigators	
Streeter and Phelps (1925) Camp (1965) Holley, et al. (1970)	$\frac{dC_B}{dt} = - k_B C_B$ C_B = BOD concentration k_B = BOD reaction rate constant
O'Connor (1967)	$\frac{dC_B}{C_B} = \frac{dQ}{Q(x)} - k_B \frac{A(x)}{Q} dx$ Q = fresh water flow rate A = cross-sectional area
Biguria, et al. (1969)	$\frac{dC_{B,C}}{dt} = k_{B,C}(C_{B,C}\big\|_{t=0} - C_{B,C}) \, , \quad 0 < t < t_n$ $\frac{dC_{B,N}}{dt} = k_{B,N}(C_{B,N}\big\|_{t=0} - C_{B,N}) + k_{B,C}(C_{B,C}\big\|_{t=0} - C_{B,C}) \, ,$ $t > t_n$ $C_{B,C}$ = concentration of carbonaceous BOD $C_{B,N}$ = concentration of nitrogenous BOD $k_{B,C}$ = carbonaceous BOD reaction rate constant $k_{B,N}$ = nitrogenous BOD reaction rate constant t_n = lag time until start of second reaction, day
Li (1972)	$\frac{\partial C_B}{\partial t} = D_L \frac{\partial^2 C_B}{\partial x^2} - u_{avg} \frac{\partial C_B}{\partial x} - k_B C_B$

Holley (1969)

$$\frac{\partial C_B}{\partial t} = D_L(t)\,\frac{\partial^2 C_B}{\partial x^2} - u_{avg}(t)\,\frac{\partial C_B}{\partial x} - k_B C_B$$

Dobbins (1964)

$$D_L\,\frac{d^2 C_B}{dx^2} - u_{avg}\,\frac{dC_B}{dx} - (k_B+k_s)C_B + L_B = 0$$

D_L = axial mixing coefficient

k_s = sedimentation rate constant of BOD

L_B = BOD addition due to local runoff

Dresnack and Dobbins (1968)

$$\frac{\partial C_B}{\partial t} = D_L\,\frac{\partial^2 C_B}{\partial x^2} - u_{avg}\,\frac{\partial C_B}{\partial x} - (k_B+k_s)C_B + L_B$$

Thomann (1973)

$$\frac{\partial C_B}{\partial t} = -\,u_{avg}\,\frac{\partial^2 C_B}{\partial x^2} - k_B C_B$$

$$\frac{\partial C_B}{\partial t} = D_L\,\frac{\partial^2 C_B}{\partial x^2} - u_{avg}\,\frac{\partial C_B}{\partial x} - k_B C_B + \frac{w(t)\delta(x)}{A}$$

$w(t)$ = input ratio of BOD

A = cross-sectional area

$\delta(x)$ = Dirac delta function on x

Bennett (1971)

$$\frac{\partial C_B}{\partial t} = D_L(t)\,\frac{\partial^2 C_B}{\partial x^2} - u_{avg}(t)\,\frac{\partial C_B}{\partial x} - (k_B+k_s)C_B + L_B$$

Ahlert (1971)

$$u(y,z)\,\frac{\partial C_{B,i}}{\partial x} = \frac{\partial}{\partial y}\left(\varepsilon_y\,\frac{\partial C_{B,i}}{\partial y}\right) + \frac{\partial}{\partial z}\left(\varepsilon_z\,\frac{\partial C_{B,i}}{\partial z}\right) - k_{B,i}C_{B,i}$$

with i = C, carbonaceous BOD
i = N, nitrogenous BOD
z = transverse direction
ε_z = lateral eddy diffusivity
y = vertical coordinate
ε_y = vertical eddy diffusivity

near the bottom of the channel. The physical mechanisms contributing to the dilution process will be discussed in detail.

The BOD concentration distribution created by a continuous source will be derived for the case of a uniform concentration over part of the cross-section at the inlet. The dependence of the BOD concentration on the reaction rate constant will be calculated explicitly.

ANALYSIS

The Instantaneous Source - An Initial Distribution Problem

In this section we will consider dispersion of material with initial distributions which are non-uniform over the cross-section of time-variable open channel laminar flows. The initial distribution will be described by two prescribed functions, $\psi(x)$ and $\phi(y)$, where x is the longitudinal coordinate and y is the vertical coordinate. The initial longitudinal distribution function, $\psi(x)$, can represent a slug or an instantaneous plane source; for a plane source $\psi(x)$ can be expressed as a Dirac delta function $\delta(x)$. The vertical distribution function, $\phi(y)$, accounts for the vertical non-uniformity of the initial distribution of material in the system as shown, for example, in Figure 2.

The general convective diffusion equation describing the concentration of solute, C_I, in laminar flow with no chemical reaction occurring in the system can be written as

$$\frac{\partial C_I}{\partial t} + \nabla \cdot \vec{V} C_I = \nabla \cdot \rho D \nabla \omega_I$$

where

$\vec{V}$ = vector representing mass average velocity

ρ = density of the fluid

D = molecular diffusion coefficient of solute

ω_I = mass fraction of solute

∇ = del operator

To facilitate the analysis the following general assumptions are made for the problem of dispersion of material in a straight, cross-sectionally uniform, and rectangular open-channel flow:

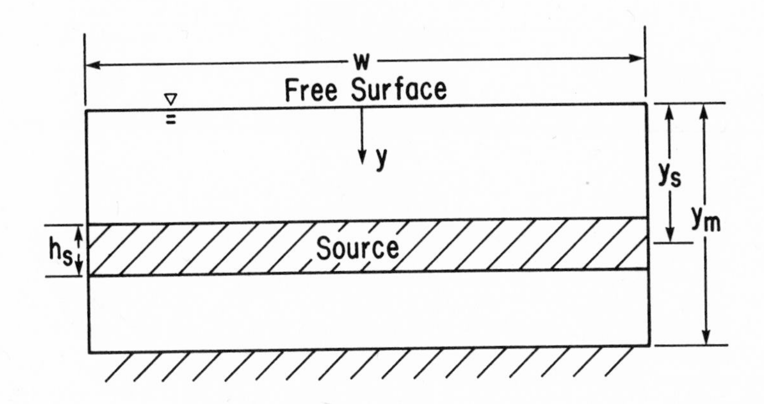

Figure 2. Sketch of the cross section of the system at $x = 0$.

(1) The transverse width, w, of the channel is much larger than its depth y_m (i.e. large aspect ratio) and there are no velocity and concentration variations in the transverse direction.

(2) The fluid is Newtonian and the flow is laminar, fully developed and time-dependent. The only non-zero velocity component is in the longitudinal direction and is described by $u(t,y)$ where t is time.

(3) The effect of mass transfer on the flow field is negligible.

(4) The flow is isothermal and the pollutant concentration is dilute so that physical properties are constant throughout the dispersion process and there is no free convection.

(5) No material transfers across the free surface or the bed of the channel.

With the given assumptions, the above convective diffusion equation can be reduced to

$$\frac{\partial C_I}{\partial t} + u(t,y)\frac{\partial C_I}{\partial x} = D\left[\frac{\partial^2 C_I}{\partial y^2} + \frac{\partial^2 C_I}{\partial x^2}\right] \tag{2}$$

along with the initial and boundary conditions:

$$C_I(0,x,y) = C_0\psi(x)\phi(y) \tag{3a}$$

$$\frac{\partial C_I}{\partial y}(t,x,0) = 0 = \frac{\partial C_I}{\partial y}(t,x,y_m) \tag{3b,c}$$

$$C_I(t,\infty,y) = 0 = \frac{\partial C_I}{\partial x}(t,\infty,y) \tag{3d,e}$$

where the subscript I denotes the concentration due to an instantaneous plane source, D is the molecular diffusion coefficient and C_0 is a reference concentration. The boundary conditions, equations (3d) and (3e), simply state that there is a finite amount of material in the system at any time. Equations (3b) and (3c) indicate that no material is transported across the free surface or into the bed.

After introducing the following dimensionless quantities:

$$\theta_I \equiv \frac{C_I}{C_0}, \quad \tau \equiv \frac{Dt}{y_m^2}, \quad X \equiv \frac{Dx}{u_0 y_m^2}, \quad Y \equiv \frac{y}{y_m}, \quad U(\tau,Y) \equiv \frac{u(t,y)}{u_0},$$

$$Pe \equiv \frac{u_0 y_m}{D},$$

the initial value problem, defined by equations (2) and (3), takes the following form

$$\frac{\partial \theta_I}{\partial \tau} + U(\tau,Y)\frac{\partial \theta_I}{\partial X} = \frac{\partial^2 \theta_I}{\partial Y^2} + \frac{1}{Pe^2}\frac{\partial^2 \theta_I}{\partial X^2} \tag{4}$$

$$\theta_I(0,X,Y) = \Psi(X)\Phi(Y) \tag{5a}$$

$$\frac{\partial \theta_I}{\partial Y}(\tau,X,0) = 0 = \frac{\partial \theta_I}{\partial Y}(\tau,X,1) \tag{5b,c}$$

$$\theta_I(\tau,\infty,Y) = 0 = \frac{\partial \theta_I}{\partial X}(\tau,\infty,Y) \tag{5d,e}$$

where u_0 is a reference velocity, Pe is the Peclet number, $\Psi(X) \equiv \psi(x)$ and $\Phi(Y) \equiv \phi(y)$.

The solution of equations (4) and (5) is formulated according to Gill (1967) as

$$\theta_I(\tau,X,Y) = \sum_{r=0}^{\infty} f_r(\tau,Y) \frac{\partial^r \theta_{Im}}{\partial X^r}(\tau,X) \tag{6}$$

where the dimensionless area-mean concentration, $\theta_{Im}(\tau,X)$, is

$$\theta_{Im}(\tau,X) \equiv \int_0^1 \theta_i(\tau,X,Y)\,dY \quad . \tag{7}$$

If equation (4) is integrated with respect to Y from Y = 0 to Y = 1, after using equations (5b,c), (6), and (7), one gets

$$\frac{\partial \theta_{Im}}{\partial \tau}(\tau,X) = \sum_{i=1}^{\infty} K_i(\tau) \frac{\partial^i \theta_{Im}}{\partial X^i}(\tau,X) \tag{8}$$

where

$$K_i(\tau) = \frac{1}{Pe^2}\delta_{i2} - \int_0^1 U(\tau,Y) f_{i-1}(\tau,Y)\,dY, \quad i = 1,2,3,\ldots \tag{9}$$

The Kronecker delta δ_{ij} is defined as

$$\delta_{ij} = \begin{cases} 1, & i = j \\ 0, & i \neq j \end{cases} \quad . \tag{10}$$

It should be noted that equation (8) is a direct consequence of the form of the solution, equation (6), which satisfies equations (4) and (5) formally. There are three main differences between this generalized dispersion model and Taylor's model (Taylor, 1953; Aris, 1956). These differences are: (1) $K_1(\tau)$ is a function of time even though the velocity field is independent of time. This occurs because of the non-uniform distribution function $\phi(y)$. For a uniform initial distribution, when material occupies the entire cross-section in a steady flow with a parabolic velocity profile, $-K_1(\tau)$ has a constant value of 2/3. This is the dimensionless mean velocity of the flow and is equal to the value in the Taylor-Aris model. (2) The coefficients $K_i(\tau)$, $i \geq 2$ also are functions of time even when the flow field, $U(\tau,Y)$ is time-independent and the initial distribution is uniform over the entire cross-section.

This occurs because the present dispersion model applies for all values of time unlike the Taylor-Aris model which is valid asymptotically for large values of time. (3) An infinite number of $K_i(\tau)$ are constructed in the present model. However, $K_1(\tau)$ and $K_2(\tau)$, in general, are the dominant terms and only small errors are incurred by neglecting all higher order terms in equation (8) if τ is not too small. For any initial distribution, as $\tau \to \infty$, $K_1(\tau)$ and $K_2(\tau)$ approach the Taylor-Aris model values, namely -2/3 and $1/Pe^2 + 8/945$ respectively.

To calculate the $K_i(\tau)$, we have to determine the equations for the f_r functions. When equation (6) is substituted into equation (4), and equation (8) is used to evaluate

$\frac{\partial \theta_{Im}}{\partial \tau}$ and $\frac{\partial^{r+1}\theta_{Im}}{\partial \tau \partial X^r}$ in terms of $\frac{\partial^i \theta_{Im}}{\partial X^i}$, by equating coefficients of $\frac{\partial^i \theta_{Im}}{\partial X^i}$,

(i=0,1,2,. . .), one can generate an infinite set of differential equations for the f_r functions given by

$$\frac{\partial f_r}{\partial \tau} = \frac{\partial^2 f_r}{\partial Y^2} - U(\tau,Y)f_{r-1}(\tau,Y) + \frac{1}{Pe^2} f_{r-2}(\tau,Y) - \sum_{i=1}^{r} K_i(\tau)f_{r-i}(\tau,Y),$$
$$r = 0,1,2,. . . \tag{11}$$

where $f_{-1} \equiv 0$ and $f_{-2} \equiv 0$.

From equations (5), (6), and (7) one can derive the necessary initial and boundary conditions for $\theta_{Im}(\tau,X)$ and the $f_r(\tau,Y)$ functions. Since

$$\theta_{Im}(0,X) = \int_0^1 \theta_I(0,X,Y)dY = \Psi(X) \int_0^1 \Phi(Y)dY \quad , \tag{12}$$

if, for convenience, one sets

$$f_r(0,Y) = 0, \quad r = 1,2,3,. . . \tag{13}$$

it follows that

$$f_0(0,Y) = \frac{\theta_I(0,X,Y)}{\theta_{Im}(0,Y)} = \frac{\Phi(Y)}{\int_0^1 \Phi(Y)dY} \tag{14}$$

The boundary conditions are

$$\frac{\partial f_r(\tau,0)}{\partial Y} = 0 = \frac{\partial f_r(\tau,1)}{\partial Y}, \quad r = 0,1,2,\ldots \tag{15a,b}$$

Also, the definition of $\theta_{Im}(\tau,X)$ results in

$$\int_0^1 f_o(\tau,Y)\,dY = 1 \tag{16a}$$

If one puts

$$\int_0^1 f_r(\tau,Y)\,dY = 0, \quad r = 1,2,3\ldots \tag{16b}$$

Furthermore, equations (5d,e) and (7) give

$$\theta_{Im}(\tau,\infty) = 0 = \frac{\partial\theta_{Im}}{\partial X}(\tau,\infty) \quad . \tag{17a,b}$$

To solve equation (8), the conditions

$$\frac{\partial^i\theta_{Im}}{\partial X^i}(\tau,\infty) = 0, \quad i = 2,3,\ldots \tag{17c}$$

may be used.

By using the defining differential equation (11), and the appropriate initial and boundary conditions, equations (13) to (16), the functions $f_r(\tau,Y)$ can be determined in principle by straightforward methods. With the $f_r(\tau,Y)$ functions known, the required coefficients $K_i(\tau)$ in the dispersion equation (8) can be obtained by using equation (9). With this information, the dispersion equation can be solved, along with its initial and boundary conditions, to provide information about the distribution of the area-average concentration as a function of time and longitudinal position. Equation (6) then can be used to obtain the local concentration distribution.

Application to Steady Laminar Flow

The theory developed above, which can accommodate time-dependent velocity fields, will be applied here to a simple steady open-channel laminar flow. The velocity profile is given by

$$U(\tau,Y) = U(Y) = 1 - Y^2 \tag{18}$$

where the reference velocity u_0 is taken to be the maximum velocity occurring at the free surface of the channel flow.

By following the previously developed procedure and applying Sturm-Liouville theory as well as Duhamel's theorem, one gets

$$f_0(\tau,Y) = 1 + \sum_{n=1}^{\infty} A_n \cos(n\pi Y) e^{-n^2\pi^2\tau} \tag{19}$$

$$K_1(\tau) = -\frac{2}{3} + 2 \sum_{n=1}^{\infty} \frac{(-1)^n A_n}{n^2\pi^2} e^{-n^2\pi^2\tau} \tag{20}$$

$$f_1(\tau,Y) = \frac{1}{6}\left(Y^2 - \frac{Y^4}{2} - \frac{7}{30}\right) + \sum_{m=1}^{\infty} \cos(m\pi Y) S_m(\tau) \tag{21}$$

$$K_2(\tau) = \frac{1}{Pe^2} + \frac{8}{945} + 2 \sum_{m=1}^{\infty} \frac{(-1)^m}{m^2\pi^2} S_m(\tau) \tag{22}$$

where

$$A_n = \frac{\int_0^1 \cos(n\pi Y) f_0(0,Y)\,dY}{\int_0^1 \cos^2(n\pi Y)\,dY} = \frac{2\int_0^1 \cos(n\pi Y)\Phi(Y)\,dY}{\int_0^1 \Phi(Y)\,dY}, \quad n = 1,2,3,\ldots \tag{23}$$

$$S_m(\tau) = -\left[\frac{4(-1)^m}{m^4\pi^4} + \frac{A_m\tau}{3}\right] e^{-m^2\pi^2\tau} + 2\sum_{n=1}^{\infty} \left\{\frac{(-1)^n A_m A_n}{n^4\pi^4}\left[e^{-(m^2+n^2)\pi^2\tau} - e^{-m^2\pi^2\tau}\right] + A_n d_{mn} \frac{e^{-n^2\pi^2\tau} - e^{-m^2\pi^2\tau}}{(m^2-n^2)\pi^2}\right\} \tag{24a}$$

and

$$d_{mn} = \begin{cases} \dfrac{2(-1)^{m+n}(m^2+n^2)}{(m^2-n^2)^2\pi^2}, & m \neq n \\[2ex] \dfrac{1}{6} + \dfrac{1}{4m^2\pi^2}, & m = n \end{cases} \tag{24b}$$

It should be emphasized here that so far we have not specified the initial distribution functions $\psi(x)$ and $\phi(y)$. Therefore, equations (19) to (24) apply to any initial distribution which is separable in the x- and y-coordinates, provided the velocity profile is described by equation (18).

As $\tau \to \infty$, irrespective of the initial distribution, it is seen from equations (19) to (24) that steady state values of f_0 and f_1 are

$$f_0(\tau\to\infty, Y) = f_{0,s} = 1 \tag{25a}$$

$$f_1(\tau\to\infty, Y) = f_{1,s}(Y) = \frac{1}{6}\left(Y^2 - \frac{Y^4}{2} - \frac{7}{30}\right) \tag{25b}$$

and $K_1(\tau)$ and $K_2(\tau)$ approach the Taylor-Aris model values which are time-independent, and are given by

$$K_1(\tau\to\infty) = K_{1,s} = -\,2/3 \tag{26a}$$

$$K_2(\tau\to\infty) = K_{2,s} = \frac{1}{Pe^2} + \frac{8}{945} \quad . \tag{26b}$$

It also can be shown that

$$K_3(\tau\to\infty) = K_{3,s} = \frac{193}{7484400} \tag{27}$$

and the higher order coefficients will decrease further in magnitude. So, $K_3(\tau)$ is more than two orders of magnitude smaller than $K_2(\tau)$ for large τ and to a good approximation all the terms involving $K_3(\tau)$ and higher order coefficients can be neglected in equation (8). The resulting truncated form of equation (8) becomes

$$\frac{\partial \theta_{Im}}{\partial \tau} = K_1(\tau)\,\frac{\partial \theta_{Im}}{\partial X} + K_2(\tau)\,\frac{\partial^2 \theta_{Im}}{\partial X^2} \tag{28}$$

which is to be solved with the initial and boundary conditions given by equations (12) and (17a,b). The transformations

$$X_1(\tau, X) = X + \int_0^{\tau} K_1(\eta)\,d\eta \tag{29a}$$

and

$$\xi(\tau) = \int_0^{\tau} K_2(\eta)\,d\eta \tag{29b}$$

reduce the problem to one of simple unsteady diffusion, where $\theta_{Im}(\xi,X_1)$ satisfies

$$\frac{\partial\theta_{Im}}{\partial\xi} = \frac{\partial^2\theta_{Im}}{\partial X_1^2} \tag{30}$$

along with

$$\theta_{Im}(0,X_1) = \Psi(X_1) \int_0^1 \Phi(Y)dY \tag{31a}$$

and

$$\theta_{Im}(\xi,\infty) = 0 = \frac{\partial\theta_{Im}}{\partial X_1}(\xi,\infty) \quad . \tag{31b}$$

Equations (30) and (31) apply to both laminar and turbulent flow but $K_1(\tau)$ and $K_2(\tau)$ are quite different in these types of flows. The solutions of equations (30) and (31) may be obtained from Crank (1956) or Carslaw and Jaeger (1959). The integrations required in equations (29a) and (29b) are simple and can be performed analytically.

A Specific Initial Distribution: Instantaneous Strip Source

Consider the case shown in Figure 2 where initially M units of material are released instantaneously into the system at the longitudinal position x = 0. The material, at time t = 0, occupies the cross-section of the channel at x = 0 as a strip h_s units thick in the vertical direction and is located with the geometric center at y_s units from the free surface; the material in the strip is uniformly distributed. This initial distribution can be represented by

$$\phi(y) = \begin{cases} 1, & y_s - \frac{h_s}{2} \leqq y \leqq y_s + \frac{h_s}{2} \\ 0, & \text{otherwise} \end{cases} \tag{32a}$$

and the reference concentration C_0 and the initial longitudinal distribution function $\psi(x)$ can be taken as

$$C_0 = \frac{M}{wy_m^2} \quad , \tag{32b}$$

$$\psi(x) = \frac{y_m^2}{h_s}\,\delta(x) \tag{32c}$$

where $\delta(x)$ is the Dirac delta function. In terms of dimensionless coordinates, equations (32a) and (32c) become

$$\Phi(Y) \equiv \phi(y) = \begin{cases} 1, & Y_s - \frac{H_s}{2} \leqq Y \leqq Y_s + \frac{H_s}{2} \\ 0, & \text{otherwise} \end{cases} \tag{33a}$$

and

$$\Psi(X) \equiv \psi(x) = \frac{\delta(X)}{H_s Pe} \quad . \tag{33b}$$

The coefficients A_n for this specific case are obtained from equation (23) as

$$A_n = \frac{4\sin(\frac{n\pi H_s}{2})\,\cos(n\pi Y_s)}{n\pi H_s} \quad , \quad n = 1,2,3,\ldots \tag{34}$$

For a uniform initial distribution, as is the case when the instantaneous source is uniform and extends vertically across the entire cross-section of the channel at $t = 0$, $H_s = 1$ and $Y_s = 1/2$ and therefore $\Phi(Y) = 1$, $(0\leqq Y\leqq 1)$. For this special case, it is interesting to note that $A_n = 0$, $(n=1,2,3,\ldots)$ and equations (19) and (20) for $f_0(\tau,Y)$ and $K_1(\tau)$ reduce to

$$f_0(\tau,Y;\, H_s=1,\, Y_s=\tfrac{1}{2}) = 1 \tag{35a}$$

$$K_1(\tau;\, H_s=1,\, Y_s=\tfrac{1}{2}) = -\frac{2}{3} \tag{35b}$$

for all τ and Y. This implies that the cross-sectional non-uniformities represented by $\Phi(Y)$ cause the time-dependence in $f_0(\tau,Y)$ and $K_1(\tau)$.

When the initial distribution functions $\Psi(X)$ and $\Phi(Y)$ are described by equations (33a) and (33b), respectively, the solution of equations (30) and (31) is given by Crank (1956) as

$$\theta_{Im}(\tau,X) = \frac{1}{2Pe\sqrt{\pi\xi(\tau)}}\, e^{-X_1^2(\tau,X)/4\xi(\tau)} \tag{36}$$

where $\xi(\tau)$ and $X_1(\tau,X)$ are given by equation (29). It should be mentioned again that equation (36) applies to both laminar and turbulent flow. However, K_1 and K_2 in equations (29a) and (29b) will be significantly different for turbulent flow.

Small τ Approximations

For small values of dimensionless time τ, the series for $K_1(\tau)$ and $K_2(\tau)$ in equations (20) and (22), respectively, converge slowly. One can take advantage of the fact that convection dominates the dispersion process at small τ and $K_1(\tau)$ and $K_2(\tau)$ can be calculated accordingly. When all the terms accounting for vertical diffusion are absent, equations (8) and (11) along with equations (13) and (14) give the following results for small τ:

$$f_0(\tau,Y) \approx f_0(0,y) = \begin{cases} \dfrac{1}{H_s}, & Y_s - \dfrac{H_s}{2} \leqq Y \leqq Y_s + \dfrac{H_s}{2} \\ 0, & \text{otherwise} \end{cases} \tag{37a}$$

$$K_1(\tau) \approx -1 + Y_s^2 + \frac{H_s^2}{12} \tag{37b}$$

$$f_1(\tau,Y) \approx \tau f_0(\tau,Y)(Y^2 - Y_s^2 - \frac{H_s^2}{12}) \tag{37c}$$

and

$$K_2(\tau) \approx \frac{1}{Pe^2} + \frac{H_s^2\tau}{3}(Y_s^2 + \frac{H_s^2}{60}) \quad . \tag{37d}$$

It also can be shown that

$$K_3(\tau) \approx \frac{H_s^2\tau^2}{6}(\frac{1}{384} + \frac{Y_s^2}{5} + \frac{H_s^2}{1260}) + \frac{\tau}{Pe^2}(1 - Y_s^2 - \frac{H_s^2}{12}) \tag{37e}$$

when τ is small. It is obvious from equations (37) that for small $\tau K_3(\tau)$ is very small in magnitude compared to $K_1(\tau)$ and $K_2(\tau)$. This is a further justification for neglecting higher order terms in the dispersion equation (8).

The Continuous Source - An Inlet Distribution Problem

Gill and Sankarasubramanian (1972a,1972b) have shown that the solution to equations (2) and (3), $C_I(t,x,y)$, can be used in conjunction with a superposition integral

$$C(t,x,y) = \int_0^t C_I(t,x,y;\lambda)\,d\lambda \tag{38}$$

to generate solutions, C(t,x,y), to a continuous source problem where the source is located at x = 0 and the source density is a function of t and y, g(t,y). One can also show that

$$u(t,y)C(t,y,x=0) = g(t,y) \quad . \tag{39}$$

Therefore, solutions to the continuous source problem can be used to solve the inlet distribution problem where the concentration at x = 0 is a specified function of t and y, C(t,y,x=0).

Gill (1975) has used this approach to study tubular reactors. Doshi et al. (1975) employed it to study reverse osmosis. Nunge and Subramanian (1977) considered atmospheric dispersion by using this method.

The Continuous Source of BOD

Subramanian et al. (1974) analyzed the problem of first-order reactions in a laminar tubular reactor by means of generalized dispersion theory. The problem of the distribution of BOD is very similar to that which they analyzed.

To obtain the local concentration of BOD, C_B, one must solve

$$\frac{\partial C_B}{\partial t} + u(t,y)\frac{\partial C_B}{\partial x} = D\frac{\partial^2 C_B}{\partial y^2} - k_B C_B \tag{40}$$

$$C_B(0,x,y) = 0 \tag{40a}$$

$$C_B(t,0,y) = C_0 \tag{40b}$$

$$\frac{\partial C_B}{\partial y}(t,x,0) = \frac{\partial C_B}{\partial y}(t,x,y_m) = 0 \tag{40c}$$

$$C_B(t,\infty,y) = \frac{\partial C_B}{\partial x}(t,\infty,y) = 0 \tag{40d}$$

Equations (40) are essentially the same as Subramanian et al. (1974) solved. First one finds the instantaneous source solution, C_{BI}, as we have done previously. Then one uses equation (38) to obtain C_B.

The main point of departure in dealing with a system with chemical reaction is the addition of $K_0\theta_{Im}$ in the dispersion model so that the lower limit on the summation in equation (8) becomes $i = 0$. Then the truncated form of equation (8) becomes

$$\frac{\partial\theta_{Im}}{\partial\tau} = K_0(\tau)\theta_{Im} + K_1(\tau)\frac{\partial\theta_{Im}}{\partial X} + K_2(\tau)\frac{\partial^2\theta_{Im}}{\partial X^2} \tag{41}$$

and the initial and boundary conditions are

$$\theta_{Im}(0,X) = \frac{m}{wy_m^2 C_0 Pe}\,\delta(X) \tag{41a}$$

$$\theta_{Im}(\tau,\infty) = \frac{\partial\theta_{Im(\tau,\infty)}}{\partial X} = 0 \tag{41b}$$

The solution of equations (41) is

$$\theta_{Im}(\tau,X) = \frac{m}{2wy_m^2 C_0 Pe\sqrt{\pi}}\;\frac{\exp[K_0\tau - \frac{X_1^2}{4\xi}]}{\sqrt{\xi}} \tag{42}$$

where $K_0 = -\gamma = -\frac{k_B y_m^2}{D}$ and X_1 and ξ are given by equations (29).

By substituting $C_0 = \frac{3m}{2wy_m u_0}$ into equation (42) and using equation (38) integrated over the area of the channel we obtain for the dimensionless area average concentration $\theta_{BM} = \frac{C_{BM}}{C_0}$ the expression

$$\theta_{Bm}(\tau,X) = \frac{1}{3\sqrt{\pi}}\int_0^{\tau}\frac{\exp\{-\gamma\tau - \frac{[X + \int_0^{\tau_1} K_1(\eta)d\eta]^2}{4\int_0^{\tau_1} K_2(\eta)d\eta}\}}{\sqrt{\int_0^{\tau_1} K_2(\eta)d\eta}}\,d\tau_1 \quad . \tag{43}$$

The asymptotic value of θ_{Im} as $\tau \to \infty$ can be calculated from equation (43). Thus,

$$\theta_{Bm}(\tau\to\infty, X) = \frac{1}{3\sqrt{\pi}} \int_0^{\infty} \frac{\exp[-\gamma\eta - \frac{(X+K_{1,s}\eta)^2}{4K_{2,s}\eta}]}{\sqrt{K_{2,s}\eta}} d\eta \tag{44}$$

where $K_{1,s} = K_1(\tau\to\infty)$ and $K_{2,s} = K_2(\tau\to\infty)$. The integration can be performed readily (Abramowitz and Stegun, 1968, p. 302) to give

$$\theta_{Bm}(\tau\to\infty, X) = -\frac{2}{3K_{1,s}\eta_1} \exp[-\frac{K_{1,s}X}{2K_{2,s}}(1-\eta_1)] \tag{45}$$

where

$$\eta_1 = [1 + \frac{4K_{2,s}\gamma}{K_{1,s}^2}]^{1/2} \tag{45a}$$

or, with the values for $K_{1,s}$ and $K_{2,s}$ from equations (26),

$$\theta_{Bm}(\infty, X) = \frac{1}{\eta_1} \exp[\frac{X}{3(\frac{1}{Pe^2} + \frac{8}{945})}(1-\eta_1)] \tag{46}$$

where

$$\eta_1 = [1 + 9(\frac{1}{Pe^2} + \frac{8}{945})\gamma]^{1/2} \quad . \tag{46a}$$

When $\gamma = 0$, the result becomes $\theta_{Bm}(\infty, X) = 1.0$.

Because of its simplicity, a common approach is to use the constant coefficient dispersion model which for this case may be written as

$$\frac{\partial\theta_{Bm}}{\partial\tau} = K_0\theta_{Bm} + K_{1,s}\frac{\partial\theta_{Bm}}{\partial X} + K_{2,s}\frac{\partial^2\theta_{Bm}}{\partial X^2} \tag{47}$$

where K_0 is equal to $-\gamma$, and $K_{1,s}$ and $K_{2,s}$ are given by equations (26). Equation (47) becomes more accurate after large values of time from the start of the process, when the coefficients in equation (41) become constants. When a step change of BOD occurs at the inlet $x = 0$, the initial and boundary conditions become

$$\theta_{Bm}(o,X) = 0 \tag{48a}$$

$$\theta_{Bm}(\tau,0) = 1 \tag{48b}$$

and

$$\theta_{Bm}(\tau,\infty) = 0 \quad . \tag{48c}$$

The solution of equations (47) and (48) can be shown to be

$$\theta_{Bm}(\tau,X) = \frac{1}{2}[\exp\{-\frac{K_{1,s}X}{2K_{2,s}}(1-\eta_1)\}\mathrm{erfc}(\frac{X+K_{1,s}\eta_1\tau}{2\sqrt{K_{2,s}\tau}}) + \exp\{-\frac{K_{1,s}X}{2K_{2,s}}(1+\eta_1)\}\mathrm{erfc}(\frac{X-K_{1,s}\eta_1}{2\sqrt{K_{2,s}\tau}})] \tag{49}$$

with η_1 given in equation (45a). At steady state, as $\tau \to \infty$, equation (49) reduces to

$$\theta_{Bm}(\tau\to\infty,X) = \exp\{\frac{-K_{1,s}X}{2K_{2,s}}(1-\eta_1)\} \quad . \tag{49a}$$

The inadequacy of equation (49), and therefore equation (47) in describing the BOD concentration distribution due to a continuous source with a partial step change at the inlet is greatest close to the source. By comparing equations (45) and (49a), one can see that they become identical only when no reaction occurs in the system.

RESULTS AND DISCUSSION

The results discussed below are for steady fully developed laminar flow with a parabolic velocity profile. It is a bit simpler to understand the mechanism of dispersion if one considers a slug, which partially occupies the cross-section as the initial configuration shown in Figure 3. A short time after the slug is introduced into the flow, longitudinal convection distorts the slug so that its forward and rear ends assume the shape of the arc of the parabola which corresponds to the velocity profile. This occurs because the particles at any level y are translated longitudinally with the velocity corresponding to that level unless molecular diffusion causes vertical movement. Now the effect of vertical diffusion of particles from the top and bottom surfaces, numbered 1 and 2, of the slug is different from that of particles

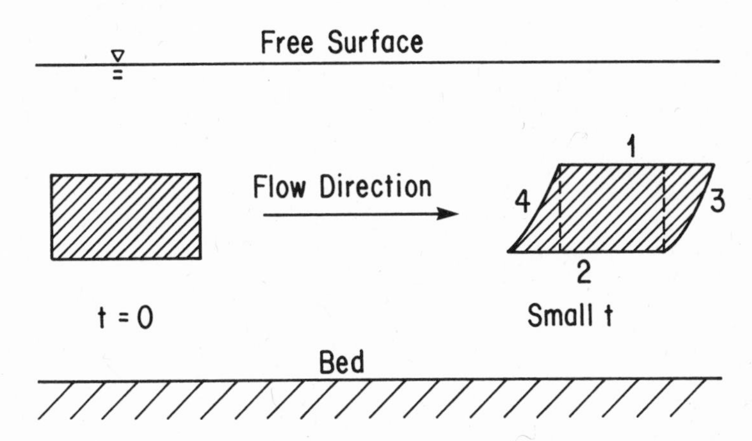

Figure 3. Sketch of a slug distorted after it is released at $t = 0$.

on the forward and rear surfaces, numbered 3 and 4. Vertical diffusion from surface 1 moves particles from slower to faster moving regions; from surface 2 it moves material from faster to slower moving regions. These effects create larger velocity differences over the region occupied by the solute and tend to reinforce the longitudinal dispersion. In contrast, vertical diffusion from surface 3 moves particles from higher to lower velocity regions; whereas from surface 4 it moves material from lower to higher velocity regions. Thus vertical diffusion from surfaces 3 and 4 causes the rear end of the slug to catch up with the front end and the net effect is to inhibit the longitudinal dispersion, or spreading out, of the solute.

In Figure 4, the negative value of the convective coefficient, $K_1(\tau)$, which plays the role of the dimensionless average velocity of the source or slug, is plotted as a function of dimensionless time, τ. By choosing the dimensionless source thickness to be $H_s = 0.05$ as a typical value, we also can see the effect of the dimen-

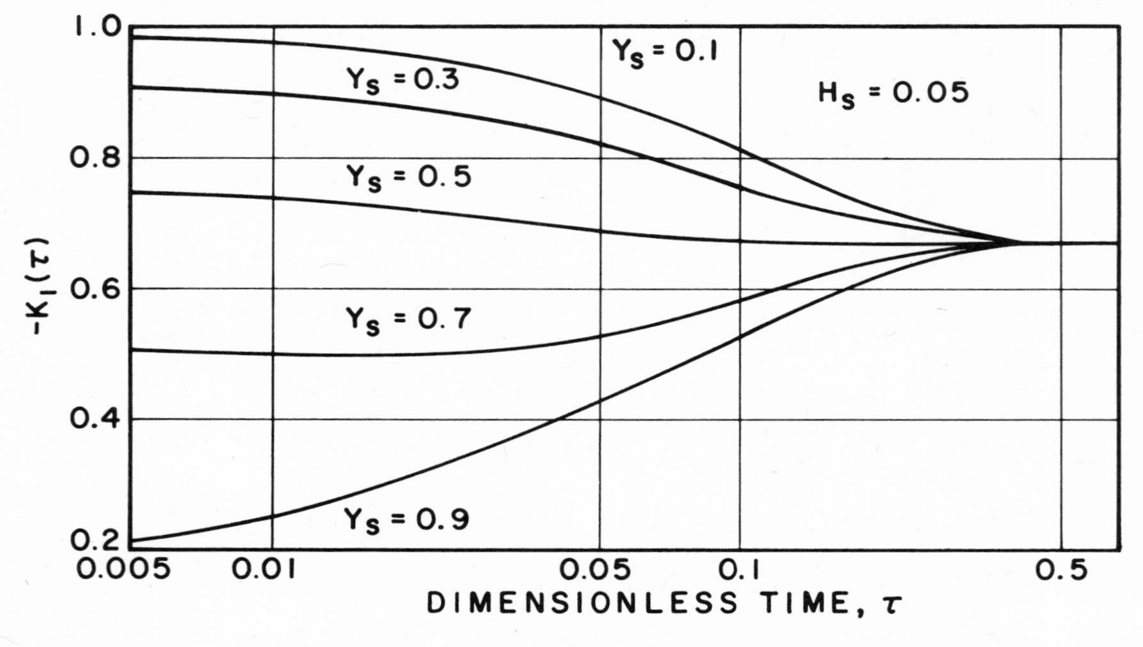

Figure 4. Plot of $-K_1$ as a function of dimensionless time, τ, with the dimensionless location of the center of the source, Y_s, as a parameter and the dimensionless thickness of the source $H_s = 0.05$.

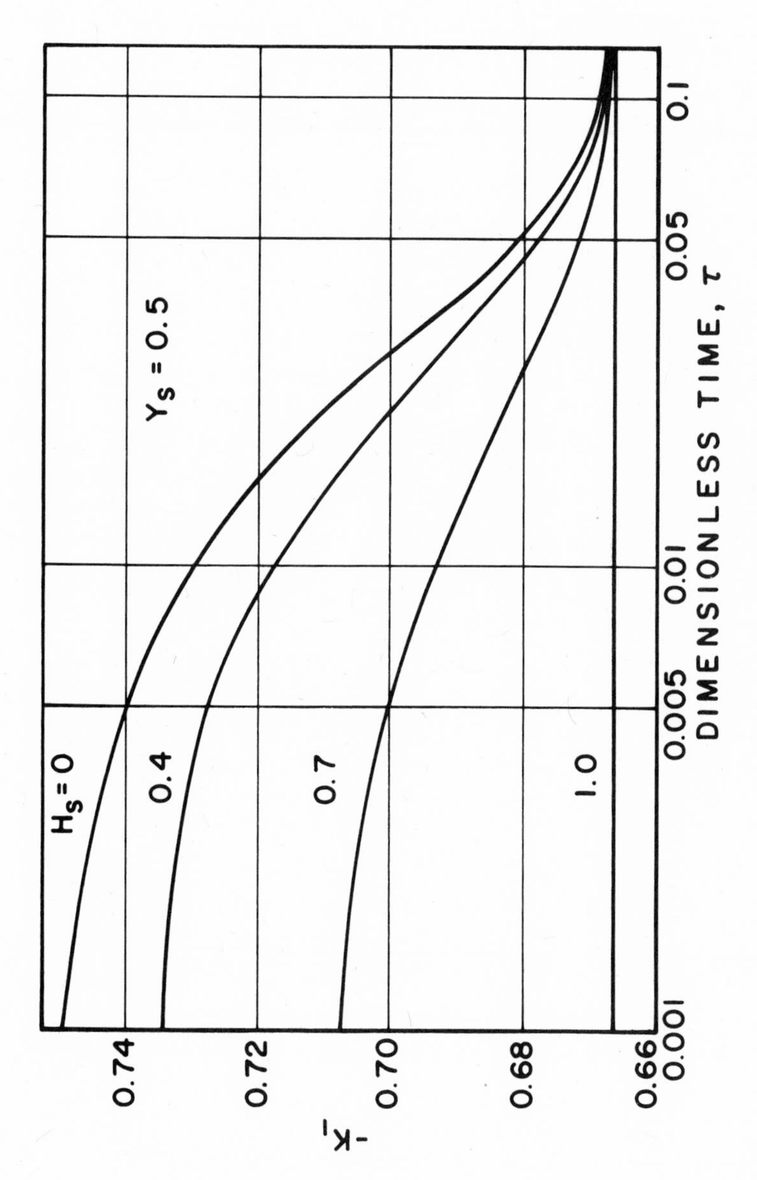

Figure 5. Plot of $-K_1$ as a function of dimensionless time, τ, with H_s as a parameter and $Y_s = 0.5$.

sionless center position of the source, Y_s, which is a parameter in the plot. After $\tau \approx 0.5$, $-K_1(\tau)$ reaches its steady state value of 2/3 for any Y_s and H_s indicating that at this moment vertical diffusion has already spread the solute over the entire cross-section of the flow and hence the average velocity of the source or slug becomes equal to that of the flow.

When the solute is located initially near the free surface, the average velocity of the source or slug is greater than the average flow velocity which it approaches as τ increases. However, for a source or slug initially close to the bottom of the channel, the situation is the opposite, and as Y_s increases, $-K_1(\tau)$ decreases for any $\tau \lesssim 0.5$.

Figure 5 is a plot of $-K_1(\tau)$ vs. τ with H_s as a parameter for Y_s arbitrarily taken as 0.5. The dimensionless average velocity of a slug of solute material at $t = 0$ can be shown easily to be $1 - Y_s^2 - H_s^2/12$. It is clear then that the initial average velocity of the solute material is decreased as the thickness of the source or slug is increased. That is, $-K_1(\tau)$ decreases as H_s increases for $\tau \lesssim 0.5$. For large τ (say > 0.5), all of the $-K_1(\tau)$ vs. τ curves converge to the same line $-K_1 = 2/3$, which is the case when the initial distribution of the solute is uniform over the entire cross-section.

Figure 6 gives the dimensionless dispersion coefficient K_2 as a function of τ with Y_s as the parameter for $H_s = 0.05$. The ordinate $K_2(\tau) - 1/Pe^2$ has been chosen instead of $K_2(\tau)$ because this removes the parametric dependence of $K_2(\tau)$ on the Peclet number, Pe, and $1/Pe^2$ represents the additive longitudinal molecular diffusion contribution. We can see that at $\tau = 0.5$, for all values of Y_s, K_2 becomes practically equal to the asymptotic value of $1/Pe^2 + 8/945$ which is same as the Taylor-Aris result. This is to be expected, since as time passes the structure of the concentration distribution of the source or slug when it is first introduced into the flow is attenuated by the convection and diffusion processes. Therefore, after the solute has been in the flow for $\tau \gtrsim 0.5$, the boundary conditions at the surface and at the bottom, rather than the initial condition, determine the behavior of the system.

It is observed that the curves for $Y_s = 0.5$, 0.7, and 0.9 intersect each other. This result may be explained as follows. Consider the pair of curves with $Y_s = 0.7$ and 0.9. At fairly small values of dimensionless time (say $\tau < 0.004$ in this example), pure convection dominates the distribution process. For the same thickness of source, the nearer to the bottom the initial position of the slug or source is, the greater the velocity variations will be in the vertical direction over the region which it occupies. Therefore, we expect to have a higher value of K_2 for $Y_s = 0.9$ than for $Y_s = 0.7$ because velocity variations in the vertical

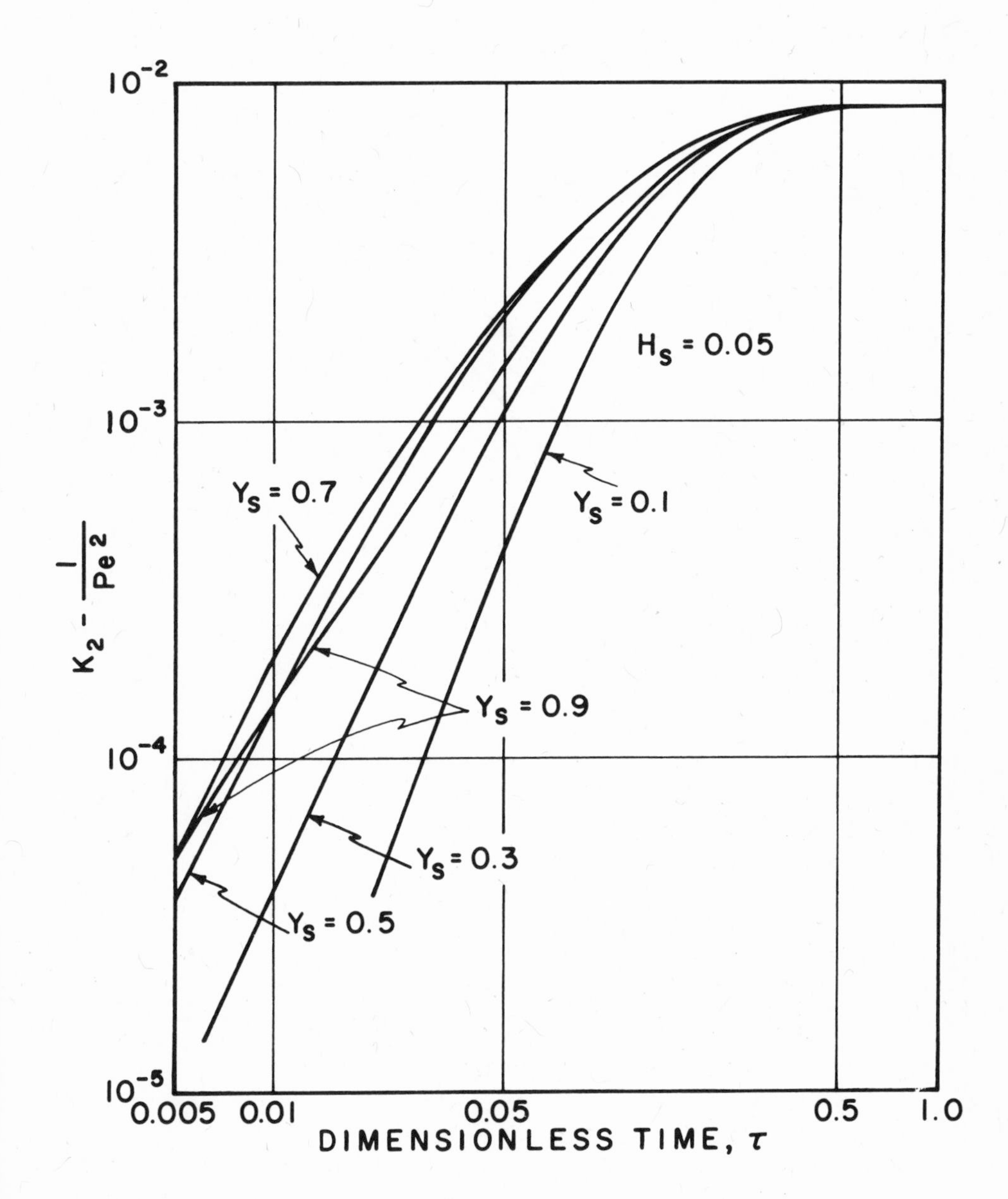

Figure 6. Plot of $K_2 - 1/Pe^2$ as a function of dimensionless time, τ, with Y_s as a parameter and $H_s = 0.05$.

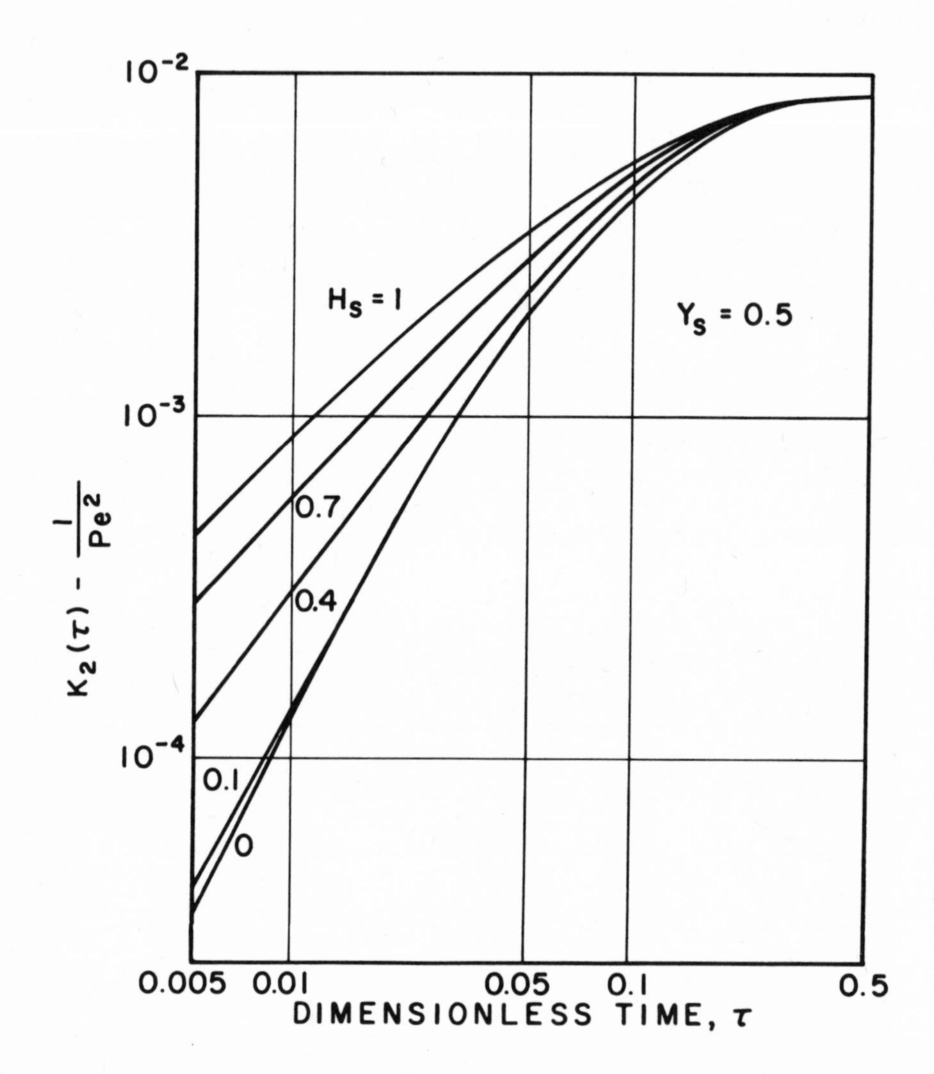

Figure 7. Plot of $K_2 - 1/Pe^2$ as a function of dimensionless time, τ, with H_s as a parameter and $Y_s = 0.5$.

direction enhance the longitudinal dispersion. When vertical diffusion starts to contribute (say τ slightly larger than 0.004 in this example), as indicated previously, vertical diffusion from surfaces 1 and 2 of Figure 3 enhance dispersion whereas from surfaces 3 and 4 it inhibits dispersion. Material transferred vertically in the initial stage of the process will have the over-all effect of aiding the longitudinal dispersion. However, the boundaries at $Y = 0$ and 1 will inhibit vertical diffusion after the process continues for some time. Compared to the case when solute initially is placed at $Y_s = 0.7$, for the same thickness of source H_s, material is reflected back from the lower boundary at an earlier stage when $Y_s = 0.9$ and this tends to reduce concentration gradients in the flow near the bed. Consequently, for a short period of time, the extent of longitudinal dispersion, as represented by the magnitude of $K_2(\tau)$, for $Y_s = 0.7$ now becomes larger than that for $Y_s = 0.9$. This reasoning also can be used to explain why $K_2(\tau)$ decreases as Y_s decreases for $\tau \lesssim 0.5$ and for $Y_s < 0.5$.

Vertical diffusion inhibits longitudinal dispersion after sufficient time has elasped for material to spread out over the entire cross-section. Finally, a so-called pseudo-equilibrium is established between longitudinal convection and longitudinal diffusion and the opposing effect of vertical diffusion. Beyond the minimum value of τ required for this to occur, K_2 maintains the same asymptotic value for all Y_s.

For a given Y_s (say 0.5), the effects of H_s on $K_2(\tau)$ can be seen from Figure 7 which includes the extreme cases of $H_s = 1$ and $H_s \to 0$. If Y_s remains the same, the velocity variations across the source become larger as H_s increases. Since the effect of velocity variations predominate in this case, $K_2(\tau)$ increases monotonically as H_s increases.

The effect of dispersion on the concentration level in response to an instantaneous source is shown by Figures 8 and 9 in which the dimensionless area-mean concentration

$$2\sqrt{\pi}Pe\theta_{Im} \quad \left(\text{or } \frac{\sqrt{\pi}Pewy_m^2}{2M} C_{Im}\right) \text{ is plotted as a function of } \tau \text{ and } X,$$

respectively, with Y_s as the parameter. One can see immediately that the concentration curve is symmetrical with respect to the dimensionless distance X; but, when breakthrough curves are plotted, a skewed time-concentration curve with a long tail results because τ appears in the denominator of the right hand side of equation (36). The velocity is larger near the free surface and, hence, the solute cloud appears sooner at the observation point for smaller Y_s as shown in Figures 8 and 9. For larger Y_s at any given τ, the spread of the source is greater and the peak mean concentration is smaller

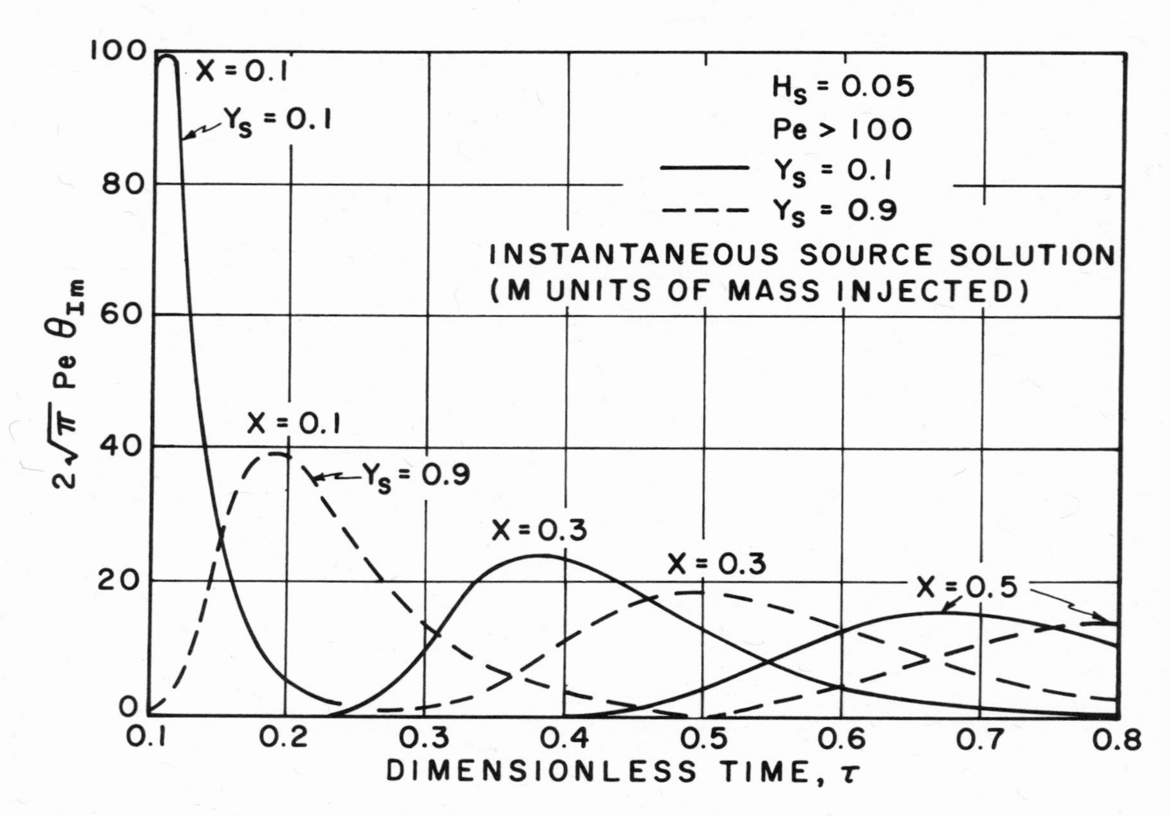

Figure 8. Plot of instantaneous source concentrations vs. dimensionless time, τ, at different dimensionless distances, X, with Y_S as a parameter (H_S=0.05 and Pe>100 in all cases).

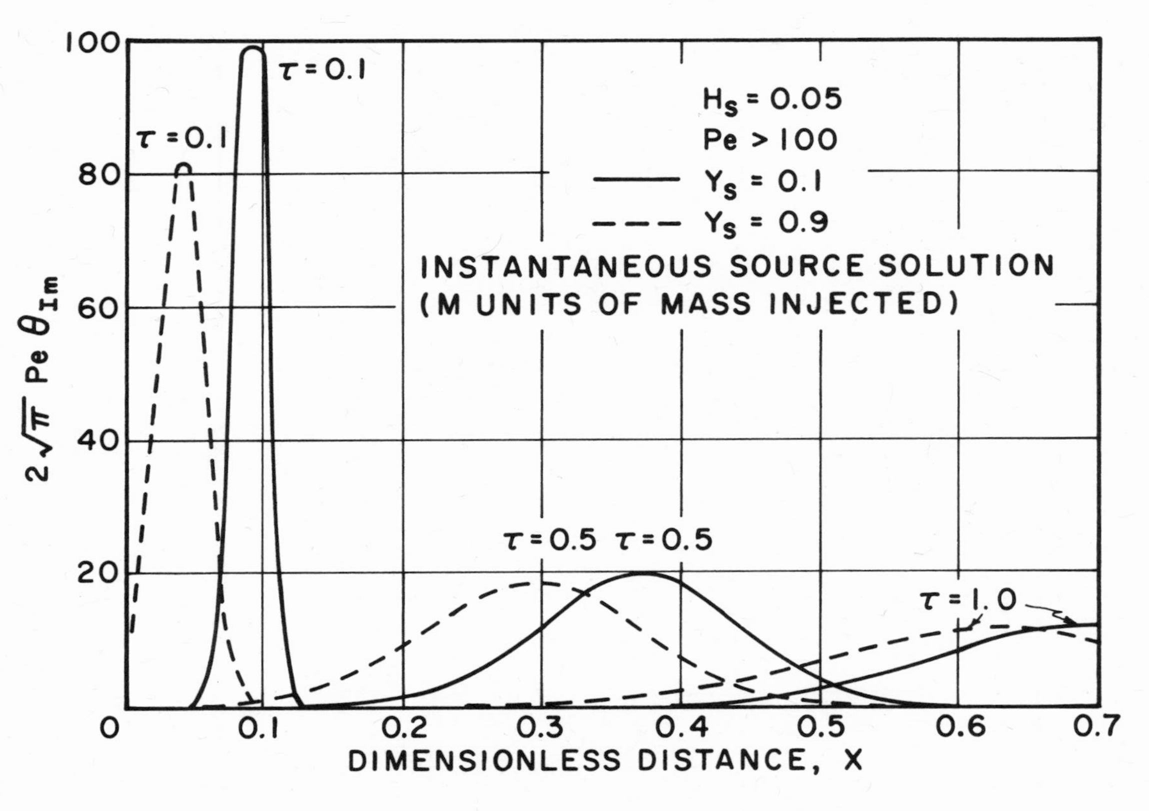

Figure 9. Plot of instantaneous source concentrations vs. dimensionless distance, X, at different dimensionless times, τ, with Y_s as a parameter (H_s=0.05 and Pe>100 in all cases).

because K_2 increases with Y_s. As the material cloud moves far enough downstream, both the spread and the peak value of the mean concentration curves will become the same for any Y_s since then K_2, for all Y_s, tends to attain the ultimate value of $1/Pe^2 + 8/945$.

Figure 10 shows breakthrough curves at X = 0.05 and 0.5 for Y_s = 0.5 and for source thickness, H_s equal 0.05 and 1. It is obvious that the spread of the curves with larger H_s is greater because of the larger velocity variations over the region which the solute occupies. Also included in Figure 10 are the breakthrough curves obtained from equation (36) with $K_1(\tau) = -2/3$ and $K_2(\tau) = 1/Pe^2 + 8/945$ which are the values of the Taylor-Aris model. It is shown clearly that significant error may be involved in predicting the concentration for small τ by using the Taylor-Aris model which applies only after the dispersion process has continued for a large enough period of time. At large τ, however, the Taylor-Aris model solution becomes close to the exact one.

Figures 11 and 12 show the breakthrough curves of BOD concentration calculated at the dimensionless longitudinal positions X = 0.01 and 1.0, respectively, for different values of the dimensionless reaction rate constant, γ. It is seen clearly that the BOD concentration is affected significantly by γ. As γ increases, the time required to achieve steady state decreases.

Figures 11 and 12 also show the effect of the source location on the BOD concentration. The effect is seen to be more significant near the source, say X = 0.01, than at a point farther downstream, say X = 1.0. With a source located near the free surface, say $Y_s = 0.1$, the BOD concentration starts to rise earlier than the one with the source near the channel bed, say $Y_s = 0.9$. This is because the BOD present near the free surface is convected downstream faster due to the higher stream velocity than the one near the channel bed. Thus, the steady state value is achieved at smaller values of τ for smaller Y_s.

Also included in Figures 11 and 12 for comparison are the results from equation (49) which is the solution of the constant-coefficient dispersion model, equation (47). Equation (49) fails to account for: (1) the effect of non-uniform distribution of the source at the inlet which is important near the source and (2) the effect of the inherent time dependence of the dispersion coefficient. Hence, equation (49) gives erroneous results in the transient period as shown in Figures 11 and 12 and its prediction is grossly in error at positions near the source. However, for large X, equation (49) can be a good approximation because the effect of the inlet condition becomes less important at a position farther downstream from the source and the coefficients, K_1 and K_2, may have achieved their asymptotic constant values before the con-

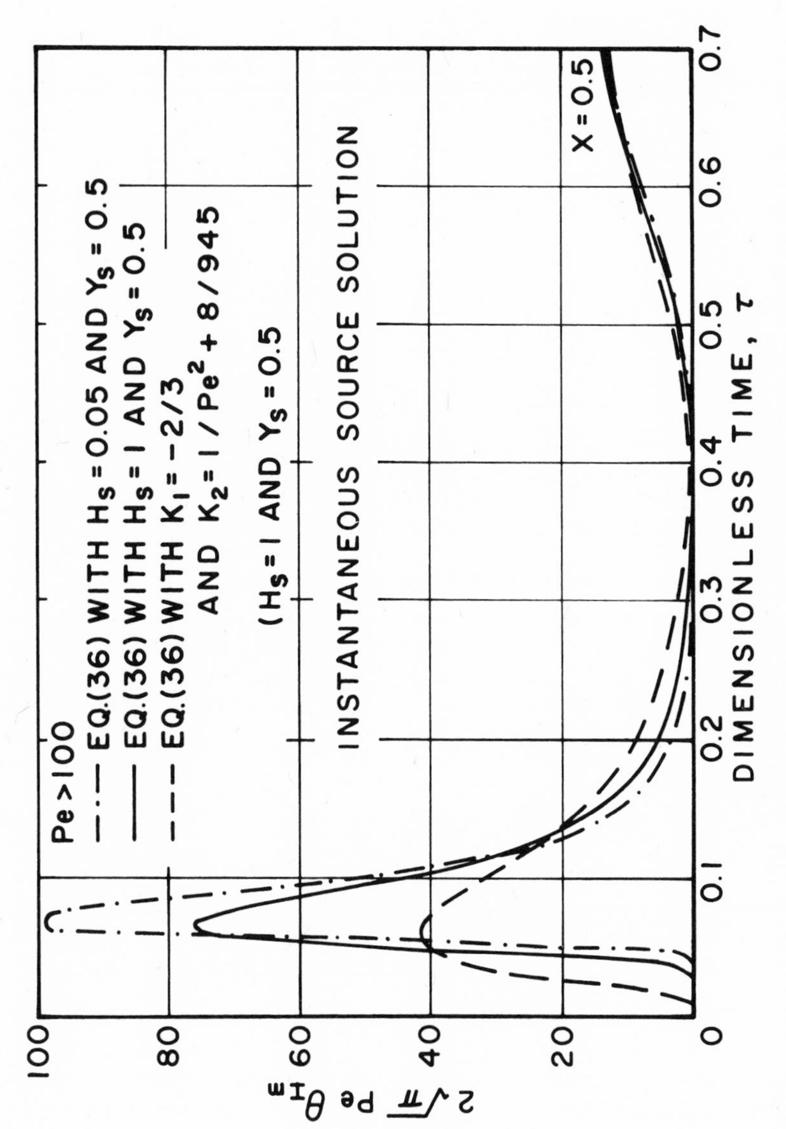

Figure 10. Dimensionless instantaneous source concentrations vs. dimensionless time, τ, at dimensionless distances X = 0.05 and 0.5 (Pe>100).

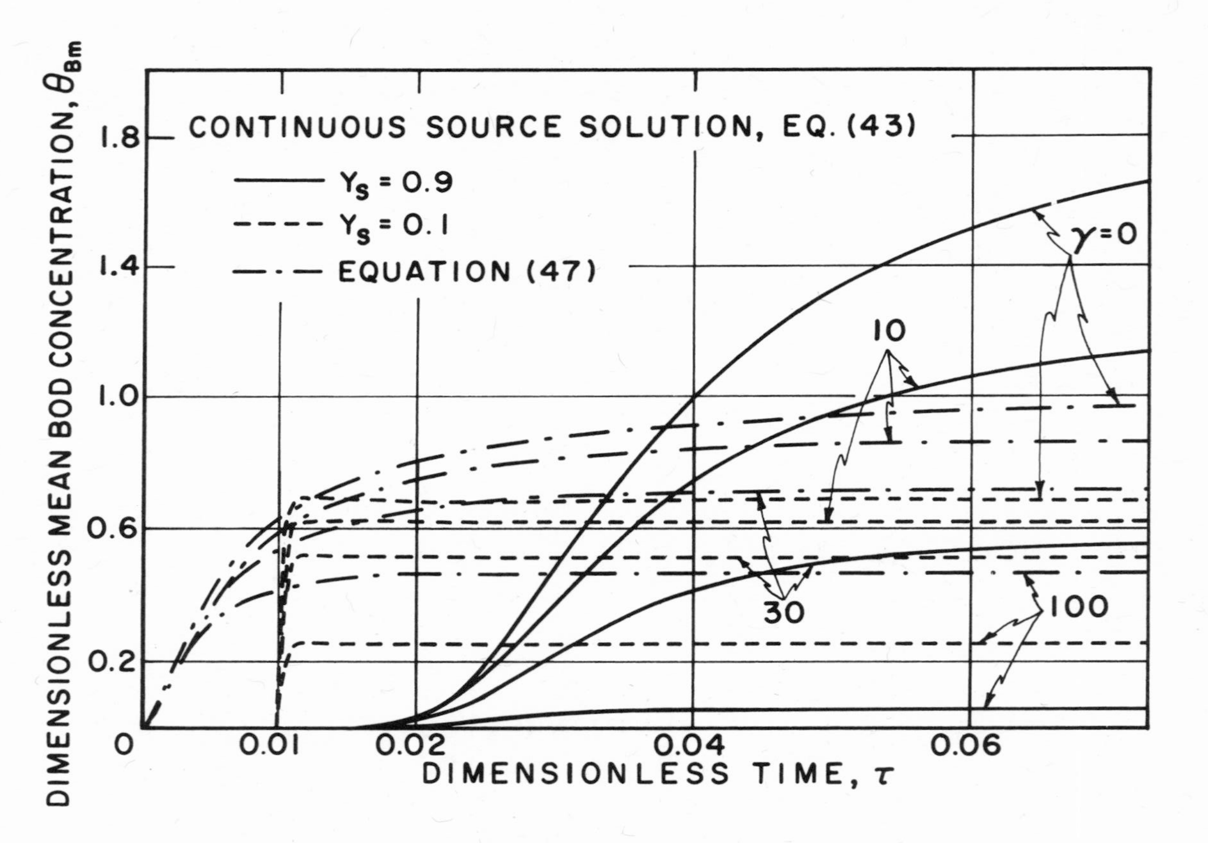

Figure 11. Breakthrough curves of BOD concentration at dimensionless longitudinal position X = 0.01; H_s = 0.1, Y_s = 0.1 and 0.9, and γ = 0, 10, 30 and 100.

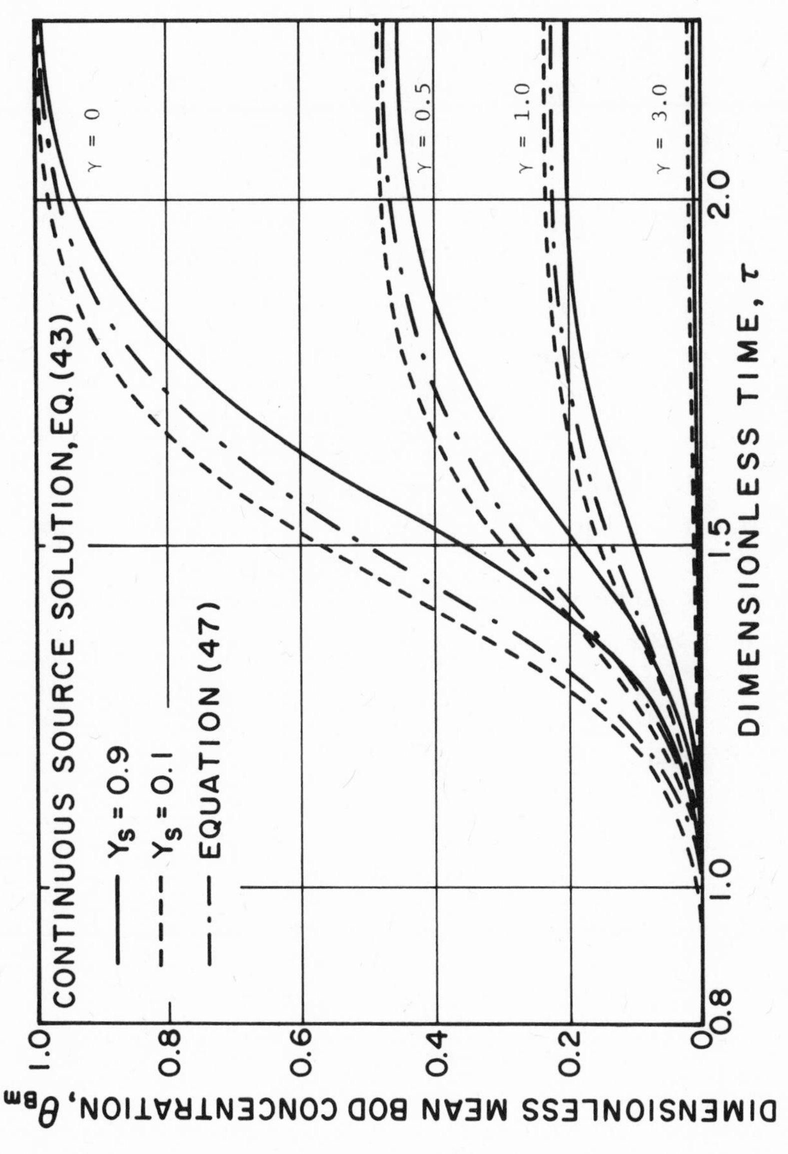

Figure 12. Breakthrough curves of BOD concentration at dimensionless longitudinal position X = 1.0; H_s = 0.1, Y_s = 0.1 and 0.9, and γ = 0, 0.5, 1 and 3.

centration at the point under consideration begins to change from its initial value to its final asymptotic value.

NOTATION

a	ratio of contact area to main stream volume
A	cross-sectional area of flow
A_n	expansion coefficients defined by equation (23)
A_r	aspect ratio of flow, width divided by depth
C	local concentration of solute
C_B	local concentration of BOD due to a continuous source
C_{Bm}	cross-sectional average of C_B
$C_{B,C}$	local concentration of carbonaceous BOD
$C_{B,N}$	local concentration of nitrogeneous BOD
C_d	concentration of solute in the dead zone
C_D	local concentration of DO
C_I	local concentration of solute due to an instantaneous source
C_m	cross-sectional average of C
C_0	bulk average concentration of solute, or reference concentration
d	ratio of contact area to dead zone volume
d_{mn}	defined by equation (24b)
D	molecular diffusion coefficient of solute
f_r	function defined by equation (11)
$f_{r,s}$	steady state value of f_r
h_s	source thickness of solute in vertical direction
H_s	dimensionless value of h_s, h_s/y_m
k_2	Taylor-Aris dispersion coefficient
$k_2(t)$	dispersion coefficient
k_B	reaction rate constant of BOD
$k_{B,C}$	reaction rate constant of carbonaceous BOD
$k_{B,N}$	reaction rate constant of nitrogenous BOD
k_s	sedimentation rate constant of BOD
K	mass transfer coefficient between the main stream and

	the dead zone
$K_1(\tau)$	dimensionless convective coefficient, u_{avg}/u_{max}
$K_{1,s}$	time asymptotic value of K_1
$K_2(\tau)$	dimensionless dispersion coefficient, $k_2(t)D/y_m^2u_0^2$
$K_{2,s}$	time asymptotic value of K_2
$K_i(\tau)$	coefficient defined by equation (9) for $i \geqq 1$
$K_{i,s}$	time asymptotic value of K_i
M	strength of instantaneous source shown in equation (32b)
Pe	Peclet number, y_mu_0/D
Q	water flow rate
S_m	function defined by equation (24a)
t	time
u	longitudinal velocity component of fluid
u_{avg}	cross-sectional average velocity of flow
u_{max}	maximum value of u at the free surface
u_0	reference velocity
U	dimensionless longitudinal velocity, u/u_0
$\vec{V}$	vector representing mass average velocity
w	width of channel
x	longitudinal coordinate
X	dimensionless longitudinal coordinate, $Dx/y_m^2u_0$
X_1	dimensionless longitudinal distance defined by equation (29a)
y	vertical coordinate
y_m	depth of flow
y_s	vertical distance of center of source
Y	dimensionless vertical coordinate, y/y_m
Y_s	dimensionless value of y_s, y_s/y_m
z	transverse coordinate

Greek Letters

γ	dimensionless reaction rate constant of BOD, $k_By_m^2/D$
$\delta(x)$	Dirac delta function of x
$\delta(X)$	Dirac delta function of X

δ_{ij} Kronecker delta defined by equation (10)

η integration variable

θ_b dimensionless local concentration of solute, C/C_0

θ_{bm} dimensionless value of C_b, C_b/C_0

θ_B dimensionless value of C_B, C_B/C_0

θ_I dimensionless value of C_I, C_I/C_0

θ_{Im} cross-sectional average of θ_I, equation (7)

θ_m dimensionless value of C_m

λ time

ξ defined by equation (29b)

ρ fluid density

τ dimensionless time, Dt/y_m^2

ω_I mass fraction of solute

Ω dimensionless quantity, $n\pi Y_s$

∇ del operator

REFERENCES

Ahlert, R. C., "Mathematical Description of Biological and Physical Processes in Heated Streams," 34a, Symp. on Biological Kinetics and Ecological Modelling, Part I. 66th AIChE National Meeting, Cincinnati, Ohio, May (1971).

Ananthakrishnan, V., W. N. Gill, and A. J. Barduhn, "Laminar Dispersion in Capillaries: Part I. Mathematical Analysis," AIChE J. 11, No. 6, 1063 (1965).

Aris, R., "On the Dispersion of a Solute in a Fluid Flowing through a Tube," Proc. Roy. Soc. Lond. A235, 67 (1956).

Bennett, J. P., "Convection Approach to the Solution for the Dissolved Oxygen Balance Equation in a Stream," Water Resources Research 7, No. 3, 580 (1971).

Biguria, G., R. C. Ahlert, and M. Schlanger, "Distributed Parameter Model of Thermal Effects in Rivers," Chem. Eng. Progr. Symp. Series 65, No. 97, 86 (1969).

Camp, T. R., "Field Estimates of Oxygen Balance Parameters," J. Sanit. Eng. Div., ASCE 91, SA5, 1 (1965).

Carslaw, H. S. and J. C. Jaeger, Conduction of Heat in Solids, 2nd Ed., Oxford University Press, London (1959).

Crank, J., The Mathematics of Diffusion, Oxford University Press, London (1956).

Davies, E. J., Personal communication.

Dobbins, W. E., "BOD and Oxygen Relationships in Streams," J. Sanit. Eng. Div., ASCE 90, SA3, 53 (1964).

Doshi, M. R., W. N. Gill, and R. S. Subramanian, "Unsteady Reverse Osmosis in a Tube," Chem. Eng. Sci. 30, 1467 (1975).

Dresnack, R. and W. E. Dobbins, "Numerical Analysis of BOD and DO Profiles," J. Sanit. Eng. Div., ASCE 94, SA5, 789 (1968).

Elder, J. W., "The Dispersion of Marked Fluid in Turbulent Shear Flow," J. Fluid Mechanics 5, 544 (1959).

Fischer, H. B., "Longitudinal Dispersion in Laboratory and Natural Streams," Ph.D. Dissertation, California Institute of Technology, Pasadena, California (1966).

Gill, W. N., "How to Solve Some Partial Differential Equations," CE Refresher Series, Chem. Eng. 69, No. 14, 145 (1962).

Gill, W. N., "A Note on the Solution of Transient Dispersion Problems," Proc. Roy. Soc. Lond. A298, 335 (1967).

Gill, W. N., "Unsteady Tubular Reactors-Time Variable Flow and Inlet Conditions," Chem. Eng. Sci. 30, 1123 (1975).

Gill, W. N. and R. Sankarasubramanian, "Exact Analysis of Unsteady Convective Diffusion," Proc. Roy. Soc. Lond. A316, 341 (1970).

Gill, W. N. and R. Sankarasubramanian, "Dispersion of a Non-Uniform Slug in Time-Dependent Flow," Proc. Roy. Soc. Lond. A322, 101 (1971).

Gill, W. N. and R. Sankarasubramanian, "Dispersion of Non-Uniformly Distributed Time-Variable Continuous Sources in Time-Dependent Flow," Proc. Roy. Soc. Lond. A327, 191 (1972a).

Glover, R. E., "Dispersion of Dissolved or Suspended Materials in Flowing Streams," USGS Prof. Paper, 433-B (1964).

Godfrey, R. G. and B. J. Frederick, "Dispersion in Natural Streams," USGS Open File Report (1963).

Hays, J. R., "Mass Transport Mechanisms in Open Channel Flow," Ph.J. Dissertation, Vanderbilt University, Nashville, Tennessee (1966).

Holley, E. R., Discussion on "Difference Modelling of Stream Pollution" by D. A. Bella and W. E. Dobbins, J. Sanit. Eng. Div., ASCE 95, SA5, 968 (1969).

Holley, E. R., F. W. Sollo, T. Micka and H. Pazwash, "Effects of Oxygen Demand on Surface Reaeration," WRC Research Report 46, University of Illinois, Urbana, Illinois (1970).

Hsieh, H., "Dispersion in Open Channel Laminar Flows," M.S. Thesis, Clarkson College of Technology, Potsdam, New York (1971).

Li, Wen-Hsiung, "Effects of Dispersion on DO-Sag in Uniform Flow," J. Sanit. Eng. Div., ASCE 98, AS1, 169 (1972).

Nunge, R. J., "Application of Exact Analytical Solution for Unsteady Advective-Diffusion to Dispersion in the Atmosphere," Atmos. Environ. 8, 984 (1974).

Nunge, F. J. and R. S. Subramanian, "Atmospheric Dispersion of Gaseous Pollutants from Continuous Source-A Model of an Industrial City," AIChE Symposium Series 73, No. 165, 10 (1977).

O'Connor, D. J., "The Temporal and Spatial Distribution of Dissolved Oxygen in Streams," Water Resources Research 3, No. 1, 65 (1967).

Patterson, C. C. and E. F. Gloyna, "Dispersion Measurements in Open Channels," J. Sanit. Eng. Div., ASCE 91, SA3, 17 (1965).

Sankarasubramanian, R. and W. N. Gill, "Dispersion from a Prescribed Concentration Distribution in Time Variable Flow," Proc. Roy. Soc. Lond. A329, 479 (1972b).

Sayre, W. W., "Dispersion of Mass in Open-Channel Flow," Ph.D. Dissertation, Colorado State University, Fort Collins, Colorado (1967).

Streeter, H. W. and E. B. Phelps, "A Study of the Pollution and Natural Purification of the Ohio River-III. Factors Concerned in the Phenomena of Oxidation and Reaeration," Public Health Bulletin 146, U.S. Dept. of HEW, Washington, DC (1925).

Taylor, G. I., "Dispersion of Soluble Matter in Solvent Flowing Slowly Through a Tube," Proc. Roy. Soc. Lond. A219, 186 (1953).

Taylor, G. I., "The Dispersion of Matter in Turbulent Flow through a Pipe," Proc. Roy. Soc. Lond. A223, 446 (1954).

Thackston, E. L. and P. A. Krenkel, "Longitudinal Mixing in Natural Streams," J. Sanit. Eng. Div. ASCE 93, SA5, 67 (1967).

Thomann, R. V., "Effect of Longitudinal Dispersion on Dynamic Water Quality Response of Streams and Rivers," Water Resources Research 9, No. 2, 355 (1973).

Thomas, I. E., "Dispersion in Open-Channel Flow," Ph.D. Dissertation, Northwestern University, Evanston, Illinois (1958).

Turner, G. A., "The Flow-Structure in Packed Beds. A Theoretical Investigation Utilizing Frequency Response," Chem. Eng. Sci. 7, 156 (1958).

Yotsukura, N. and M. B. Fiering, "Numerical Solution to a Dispersion Equation," J. Hyd. Div., ASCE 90, HY5, 83 (1964).

ACKNOWLEDGEMENT

This work was supported in part by National Science Foundation Grant KO34380.

MOMENTUM TRANSFER AT THE AIR-WATER INTERFACE

Omar H. Shemdin [1]

Jet Propulsion Laboratory

California Institute of Technology

ABSTRACT

Wind action over water generates waves and surface drift. The logarithmic velocity profile in air can be used to determine the surface stress with a suitable choice of roughness height. A corresponding logarithmic profile is found below the interface and yields a surface stress equivalent to that found from the air profile. The wind induced set-up depends on water depth, surface and bottom boundary stresses, and atmospheric pressure gradient. Modelling relationships to simulate set-up under laboratory conditions are discussed. The momentum transfer from wind to waves is found to be approximately 10-20 percent of the total wind induced stress. The net energy transfer to any wave spectral component depends on nonlinear wave-wave transfers in addition to that received directly from wind less that dissipated by viscosity or turbulence.

INTRODUCTION

Wind action over water generates waves and surface drift. Waves contribute to surface drift through Stokes mass transport. The transfer of momentum from air to water occurs partially through normal stress which contributes primarily to surface drift. Details of momentum and energy transfers from air to water are not totally understood. The extensive research in progress is aimed at filling existing gaps. The research on wave generation is devoted to developing models consistent with wave physics in contrast to empirical

---

[1] Professor, University of Florida, on leave of absence.

procedures. The generation of current by wind and its prediction is demanding increased attention. Wind induced circulation in shallow water plays an important role in flushing of waterways and dispersion of heat and pollutants introduced into the waterways.

Modelling of wind over water is useful in arriving at solutions to engineering problems where wind plays a dominant role in water circulation. Use of wind models is feasible provided that scale of modelled area is small to justify neglecting effect of earth rotation. The available literature on wind and wind driven circulation is primarily devoted to understanding properties of shear flows in air and induced motion in water. An abundance of field and laboratory results are reported but only a few reported studies provide insight relevant to modelling of wind over water.

WIND PROFILE OVER WATER

The wind profile, $U_a(z)$, is most commonly assumed to have the logarithmic distribution

$$U_a(z) = \frac{U_{*a}}{\kappa} \ln \frac{z}{z_{oa}} \quad , \tag{1}$$

where κ is the Von Karman constant, z_{oa} is the roughness height, and U_{*a} is the friction velocity ($= \sqrt{\tau_s/\rho_a}$). τ_s is the surface stress and ρ_a is the air density. Such a profile is not uniformly endorsed by all investigators. De Leonibus (1971) reported that he seldom encountered a logarithmic distribution in his field measurements of the wind profile with four cup anemometers. Ruggles (1969), however, indicated that 90% of his field measurements of the wind profile followed the logarithmic distribution. In the laboratory, Lai and Shemdin (1971) reported that the wind profile can be approximated by a logarithmic distribution. The profiles are sensitive to the presence of waves, however, which induce an increase in U_{*a} as shown in Fig. 1. The influence of waves on U_{*a} and the drag coefficient was also reported by De Leonibus (1971) by comparing spectra of waves and turbulence above the water surface. Other evidence of the wave-induced organized motion was reported by Elder, Harris and Taylor (1970) in the field and by Lai and Shemdin (1971) in the laboratory.

The drag coefficient , C_z, of wind is defined

$$\tau_s = C_z \rho_a U_a^{\,2}(z) \quad , \tag{2}$$

or in terms of shear velocity, U_{*a}, the drag coefficient is expressed

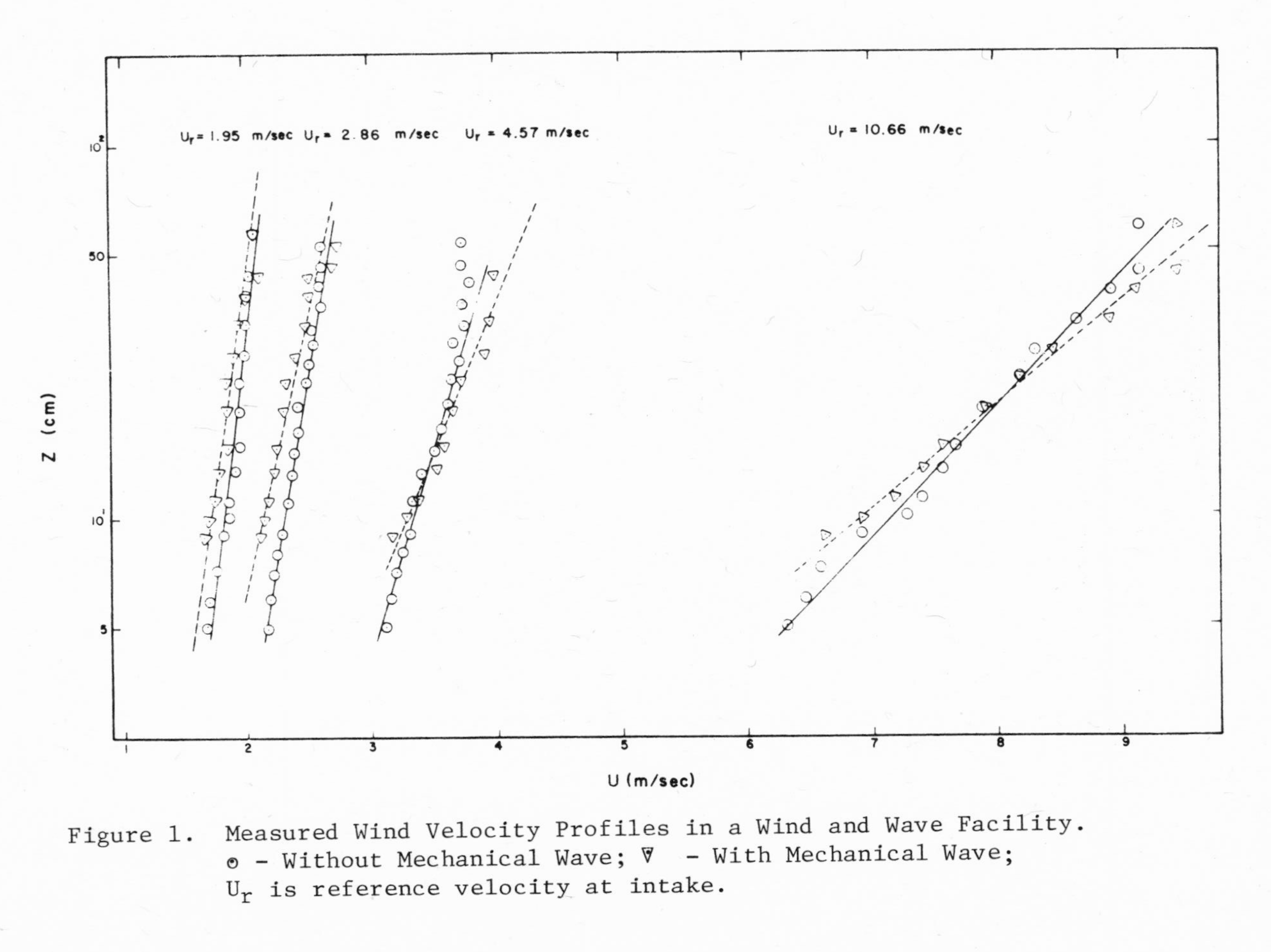

Figure 1. Measured Wind Velocity Profiles in a Wind and Wave Facility. ⊙ – Without Mechanical Wave; ▽ – With Mechanical Wave; U_r is reference velocity at intake.

$$C_z = \left(\frac{U_{*a}}{U_a(z)}\right)^2 \quad . \tag{3}$$

The drag coefficient can be evaluated by direct measurement of the turbulent flux, by the profile method assuming logarithmic distribution, and by the eddy viscosity method. The values of drag coefficient reported under both field and laboratory conditions show considerable scatter. Comprehensive reviews have been given by Wilson (1960) and Wu (1969). The scatter is greater for light winds and can be attributed to variations in air stability, sea state and the method used to evaluate the drag coefficient.

If the logarithmic profile given in Eq. 1 is used, the drag coefficient becomes

$$C_z^{-1/2} = \frac{1}{\kappa} \ln\left(1/C_z \; F^2\left(\frac{gz_o}{U_{*a}^{\;2}}\right)\right) \quad , \tag{4}$$

where $F^2 = U_z^2/gz$. Charnock suggested the ratio $gz_o/U_{*a}^{\;2}$ is a constant based on field measurements and Wu (1969) confirmed this suggestion in his laboratory channel at a fixed station. For constant $gz_o/U_{*a}^{\;2}$ the drag coefficient becomes a function of the Froude number and convenient elevations can be selected for field and laboratory measurement to facilitate modelling as suggested by Wu (1971). A substantial amount of field evidence reported by Roll (1965) and Kitaigarodski and Volkov (1965), however, indicates that $gz_o/U_{*a}^{\;2}$ is not a constant as shown in Fig. 2. A review by Shemdin and Mehta (1972) of results obtained in both the Stanford University and Colorado State University wind and waves facilities also suggests that $gz_o/U_{*a}^{\;2}$ is not a constant but depends on both fetch and wind speed; consequently a simple Froude model cannot be easily supported. A demonstration of the above is given in Fig. 3 where the drag coefficient C_{10} is graphed versus the wave Reynolds number, $c\sigma/\nu_\omega$, where C is the wave phase speed, σ is the rms wave energy and ν_ω is the kinematic viscosity of water.

WIND-INDUCED DRIFT

Of the many reported studies on air-sea interaction only a few give detailed results of the wind-induced drift. In the field, Van Dorn (1953) investigated set-up and surface drift in a pond under the action of the wind. He found the surface drift to remain constant even after spreading detergent over the water. Keulegan (1951) found the surface drift to be 3% of the wind speed. The surface drift is reported to vary from 2.4% to 5%. The 3% value appears to be a reasonable average.

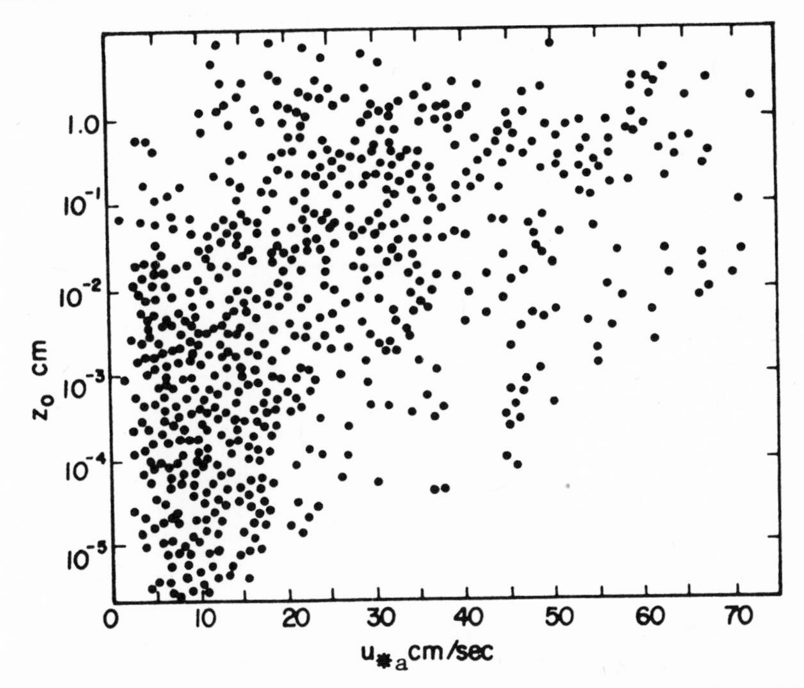

Figure 2. z_o vs. U_{*a} (after Kitaigorodsky and Volkov, 1965).

Using neutrally buoyant discs at the surface and over a sufficient depth below the surface in a laboratory facility Shemdin (1972) proposed the following wind-induced current profile based on measurements shown in Fig. 4,

$$U_w(o) - U_w(z) = \frac{U_{*w}}{\kappa} \ln\left(\frac{z}{z_{ow}}\right) , \qquad (5)$$

where $U_w(o)$ is the surface drift, U_{*w} is the water shear velocity and z_{ow} is the roughness height below the surface. The surface shear stress can be computed from shear velocities in air and water.

$$\tau_s = \rho_a U_{*a}^2 = \rho_w U_{*w}^2 , \qquad (6)$$

where ρ is the water density. The results are shown in Fig. 5 and suggest continuity of tangential stress across the air-water interface. The surface drift was found to be 3% of the free stream wind speed. The roughness heights were found to be of the order $z_{oa} \simeq z_{ow} \simeq$ 0.1 mm. From Eq (3) using $z_{oa} \simeq 0.1$ mm it follows

$$U_a(10) = 29\ U_{*a} , \qquad (7)$$

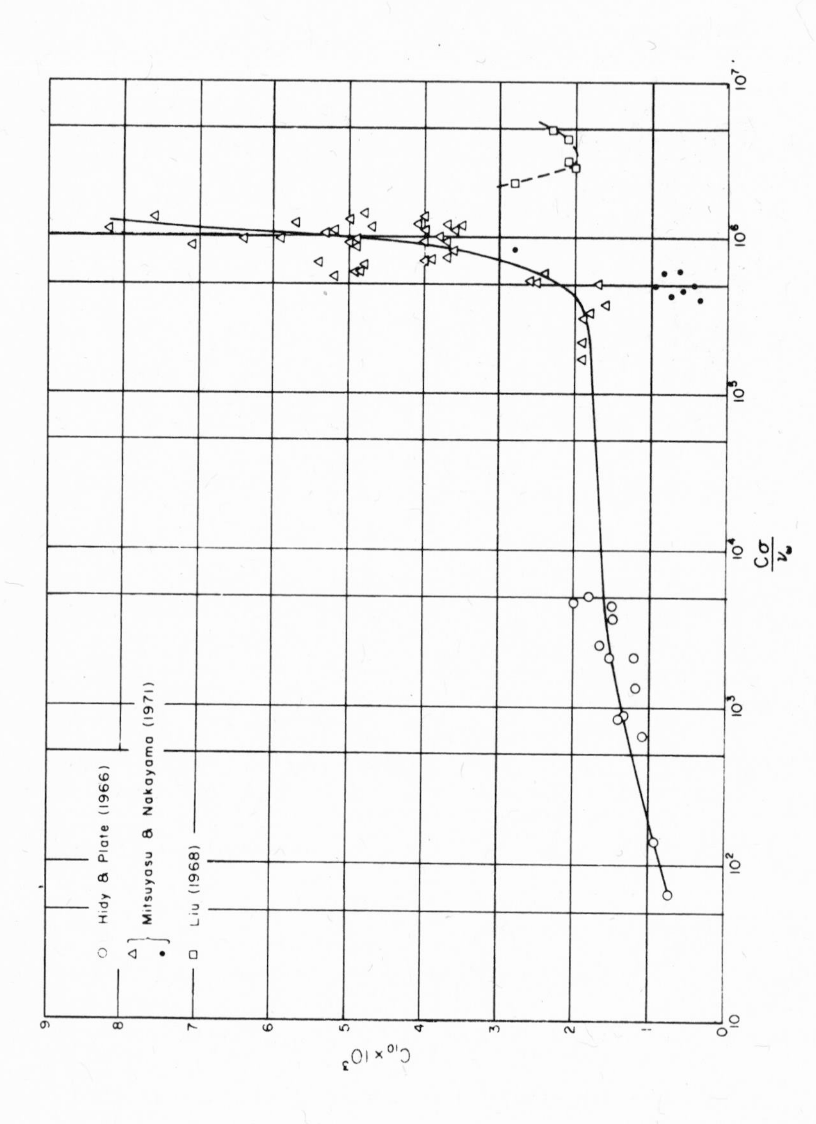

Figure 3. Dependence of Drag Coefficient on $c\sigma/\nu_\omega$.

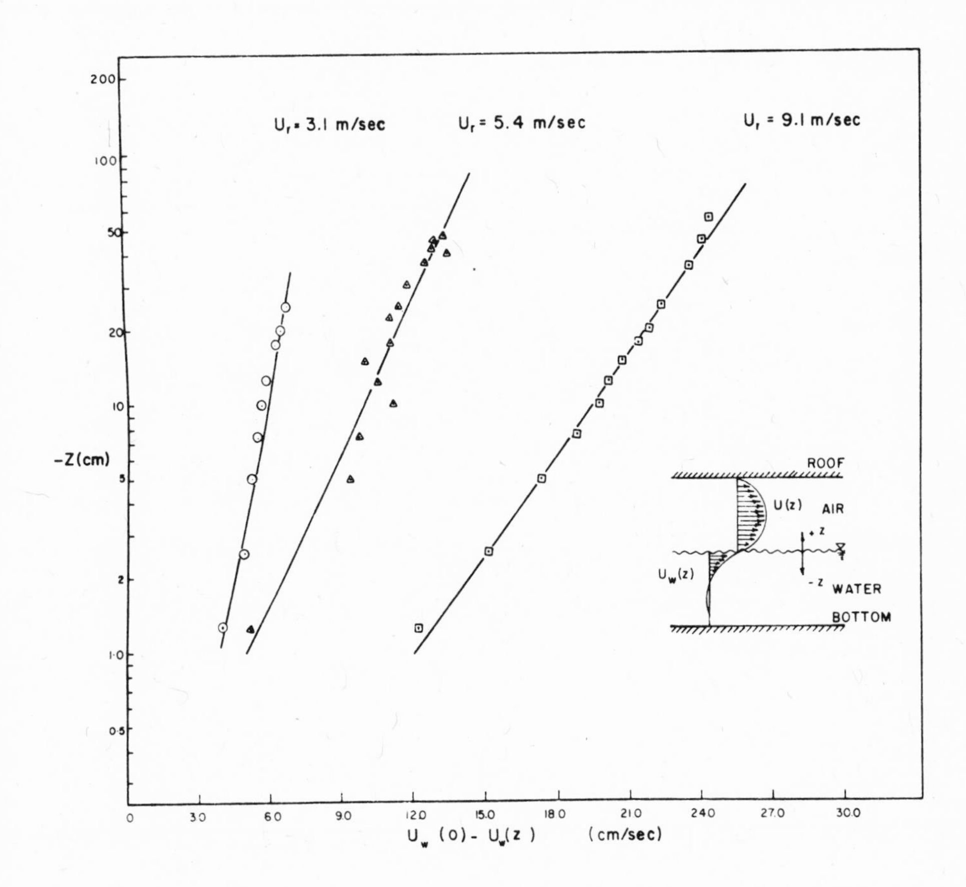

Figure 4. Logarithmic Wind-Induced Drift Profiles .

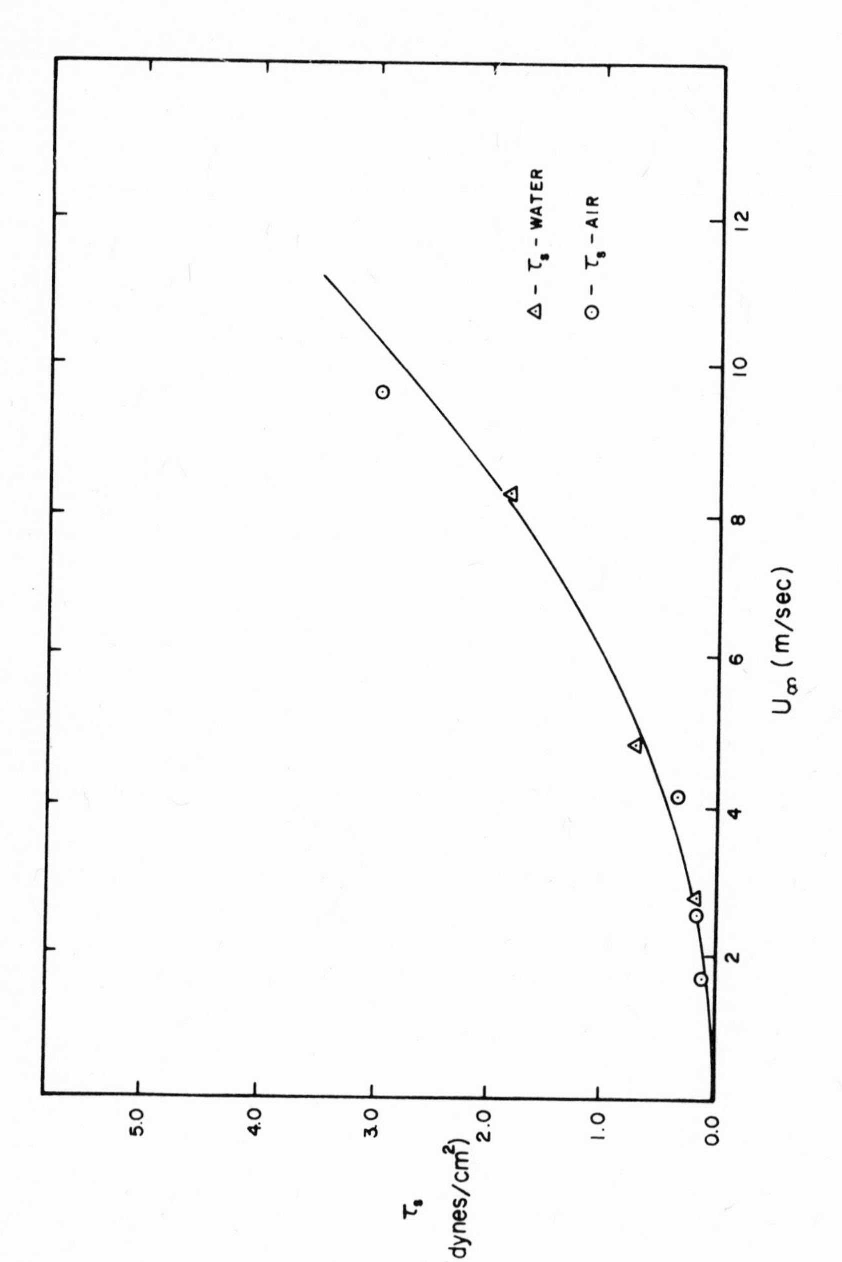

Figure 5. Equilibrium of Stress Above and Below the Interface.

where $U_a(10)$ is the wind speed at 10 m. This result is found frequently in the field. From Eq. 6 it follows

$$U_{*a} = \sqrt{\frac{\rho_w}{\rho_a}}\, U_{*w} = 29\, U_{*w} \quad , \tag{8}$$

and using Eq. 5 with $z_{ow} \simeq 0.1$ mm

$$U_w(o) - U_w(-10) = 29\, U_{*w} \quad , \tag{9}$$

where $U_w(-10)$ is the current at a depth of 10 m. It follows that

$$U_w(o) - U_w(-10) = U_{*a} = \frac{U_a(10)}{29} = 0.034\, U_a(10) \quad . \tag{10}$$

The wind-induced current at a depth of 10 meters is negligible compared to the surface drift so that the above relationships predict the surface drift to be 3.4% of the wind speed at 10 meters above the surface, a result which is consistent with laboratory measurements and with the field investigation of Van Dorn (1953).

Calculation of the surface drift based on the Stokes mass transport were made by Kenyon (1969) and Bye (1967) for a fully developed sea wave spectrum. The surface drift was found to be approximately 3% of the wind speed. A similar calculation made by Shemdin (1972) for the fetch-limited laboratory waves was found to predict a surface drift of approximately 0.3% of the wind speed while the surface drift was measured at 3% of the wind speed.

WIND-INDUCED SET-UP AND MODELLING OF WIND OVER WATER

The surface slope (set-up) is important in determining storm surge elevations. Using Fig. 6 as a defnition sketch it can be shown (see Ursell 1956)

$$\frac{\partial s}{\partial x} = \frac{\tau_s - \tau_b}{\rho_w g\,(h + s)} - \frac{1}{\rho_w g}\frac{\partial p_a}{\partial x} - \frac{1}{\rho_w h(h + s)}\frac{\partial}{\partial x}\int_{-h}^{s} \rho_w U_w^2\, dz, \tag{11}$$

where p_a is the aerodynamic static pressure, s is the surface elevation above an origin and the other variables are as defined in Fig. 6. Ursell (1956) evaluated the relative importance of the last term on the right hand side of Eq. 11 and found it to be significant for the type of apparatus and experiments conducted by Keulegan (1951). In the field where the water depth is large the contribution of this term is negligible.

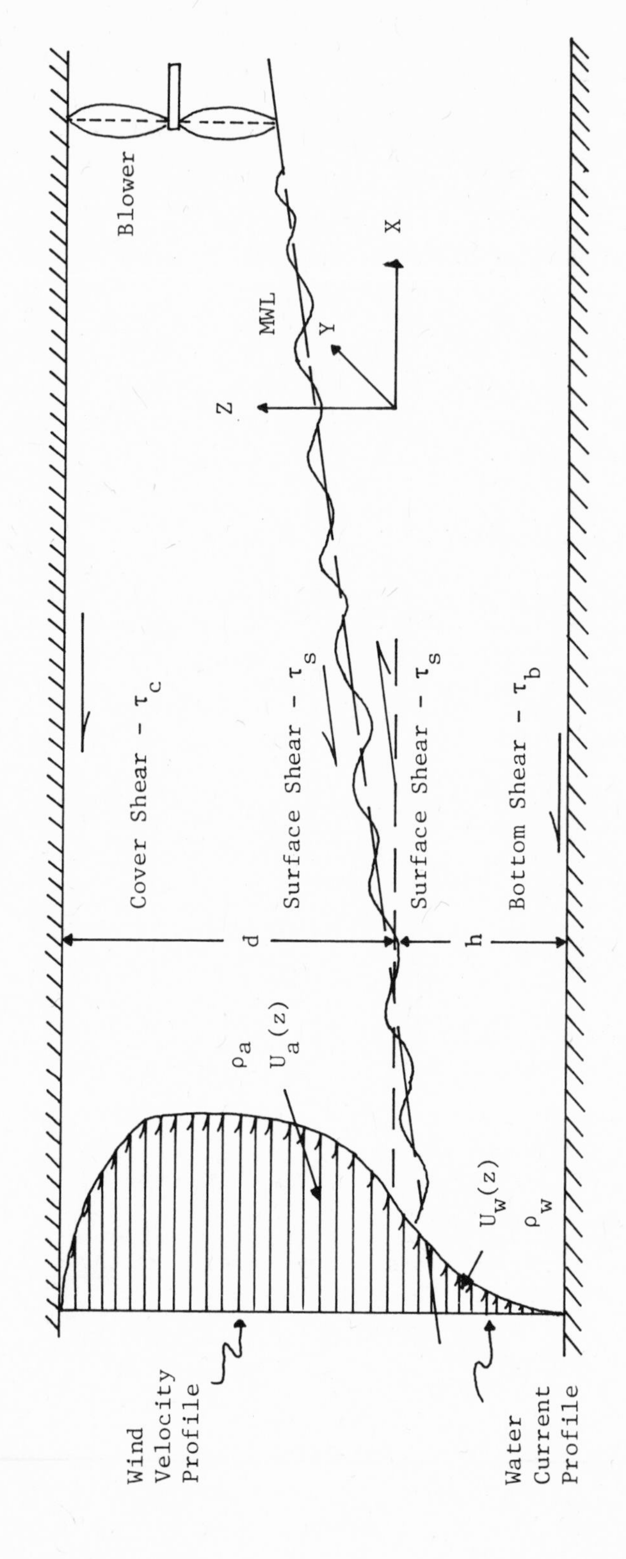

Figure 6. Definition Sketch for Wind Models.

The available field and laboratory results suggest the following guidelines for modelling wind over water:

1. A logarithmic distribution may be used to describe the wind profile.
2. The drag coefficient is sensitive to sea state and thermodynamic stability, and cannot be expressed simply in terms of a Froude number based on wind speed and elevation of wind measurement.
3. The surface drift is approximately 3% of the wind speed regardless of sea state. In an unlimited fetch situation the major portion of the drift is by the Stokes mass transport (Bye 1967; Kenyon 1969).
4. The wind-induced current profile follows a logarithmic distribution which gives a surface stress consistent with that obtained from the wind profile. Such a velocity distribution is not valid for laminar flow.
5. The momentum flux gradient in the direction of wind contributes to set-up when water depth is small.

In the following modelling relationships are derived for wind-induced circulation under steady conditions. The surface slope (Eq. 11) reduces to the simplified form when water depth is not small

$$\frac{\partial s}{\partial x} = \frac{\tau_s - \tau_b}{\rho_w g h} - \frac{1}{\rho_w g}\frac{\partial p_a}{\partial x} , \tag{12}$$

where the symbols are shown as in Fig. 6. It is assumed that $s \ll h$ and the last term on the right hand side of Eq. 11 is negligible. The corresponding equation for air is

$$\frac{\tau_s + \tau_c}{d} = \frac{\partial p_a}{\partial x} . \tag{13}$$

Eliminating the pressure term in Eq. 12 and 13 yields

$$\frac{\partial h}{\partial x} = \frac{(\tau_s - \tau_b)}{\rho_w g h} + \frac{(\tau_s + \tau_c)}{\rho_w g d} . \tag{14}$$

The bottom shear stress is expressed as

$$\tau_b = -n\tau_s , \tag{15}$$

where n varies from a negative value for an open basin to 0.5 for laminar flow in a closed basin. The surface shear stress is described in terms of the drag coefficient given in Eq. 2 . Similarly the roof shear stress may be described in terms of the air velocity, U_a, and a roof friction coefficient, C_c

$$\tau_c = C_c \rho_a U_a^2(z) \quad . \tag{16}$$

Equation 14 becomes

$$\frac{\partial s}{\partial x} = C_z \left(\frac{\rho_a}{\rho_w}\right) \frac{U_a^2(z)}{gh} \left[(1 + n) + \frac{h}{d} + \frac{C_c}{C_z}\frac{h}{d}\right] , \tag{17}$$

which describes the air-water motion in the model. In the prototype, d is large compared to h so that

$$\frac{\partial s}{\partial x} = C_z \left(\frac{\rho_a}{\rho_w}\right) \frac{U_a^2(z)}{gh} (1 + n) \quad . \tag{18}$$

Denoting the surface slope in the model by, S_m, and in the prototype by, S_p, and their ratio by, η_s

$$\eta_s = \frac{S_m}{S_p} , \tag{19}$$

it is shown that

$$\eta_s = \left[\frac{\eta_{c_z} \eta^2_{U_a(z)}}{\eta_h}\right] \left[\frac{(1 + n)_m + \left(\frac{h}{d}\right)_m + (C_c/C_z) \left(\frac{h}{d}\right)_m}{(1 + n)_p}\right] . \tag{20}$$

In a small scale model n_m may correspond to laminar flow and assume a value different from n_p which normally corresponds to turbulent flow.

Vertical distortion in a model may be implemented as follows:

$$\eta_s = \eta_z/\eta_x ,$$

where η_z and η_x refer to the vertical and horizontal scales, respectively. In such a model the velocity scale becomes

$$\eta^2_{U_a} = \frac{\eta_z}{\eta_x \eta_{c_z}} \left[\frac{(1 + n)_p}{(1 + n)_m + \left(\frac{h}{d}\right)_m + \frac{C_c}{C_z} \left(\frac{h}{d}\right)_m}\right] . \tag{21}$$

An appropriate procedure in modelling wind-driven circulation is to obtain field measurements of the velocity profile and to calculate C_z. Assuming that n has the same value both in model and in prototype the wind velocity in the model can be estimated and used to determine the model dimensions and to verify the n value in the model. It is noted that for laminar flow n = 0.5 and for turbulent flow n = 0.1 - 0.25 in a closed basin. The horizontal and vertical scales may be varied to obtain an appropriate model size. When the size of the model is selected and the wind scale is determined accurately from the measured velocity profile in the model the surface drift in the model is calculated as 3% of the wind speed when $(U_w(o)g/\nu_\omega > 10^4$, where ν_ω is the kinematic viscosity of water. Otherwise a smaller drift results as given by Keulegan (1951). The cover shear stress coefficient C_c is obtained from standard calculations for flows over flat plates.

MOMENTUM TRANSFER TO WAVES

Momentum transfer to waves occurs primarily through normal stress at the water surface. A turbulent shear flow perturbed by surface waves produces a normal stress which is out of phase with the surface wave such that the peak pressure is on the upwind slope of the wave. The streamline patterns for such a flow is shown in Fig. 7 based on velocity profile measurements below the wave crest obtained with a wave follower (Shemdin, 1969). The asymmetric pressure distribution along the wave profile produces momentum transfer to the wave. This is demonstrated in the most simple form by considering a propagating sinusoidal wave, η^1, defined by

$$\eta^1 = \left[a_o \exp i \left\{ k(x - ct) \right\} \right] , \quad (22)$$

where, a_o is the wave amplitude, k is the wave number, and c is the wave speed. The perturbation pressure at the water surface is denoted by

$$P_{as} = R_e \, (a + ib) \, \rho_a \, g \, \eta^1 , \quad (23)$$

where a and b are in-phase and out-of-phase pressure coefficients, ρ_a is the air density, and g is gravity. Momentum transfer to wave, M, is given by

$$M = \frac{1}{L} \int_0^L P_{as} \frac{\partial \eta^1}{\partial x} \, dx , \quad (24)$$

where L is the wave length, so that after integration it becomes

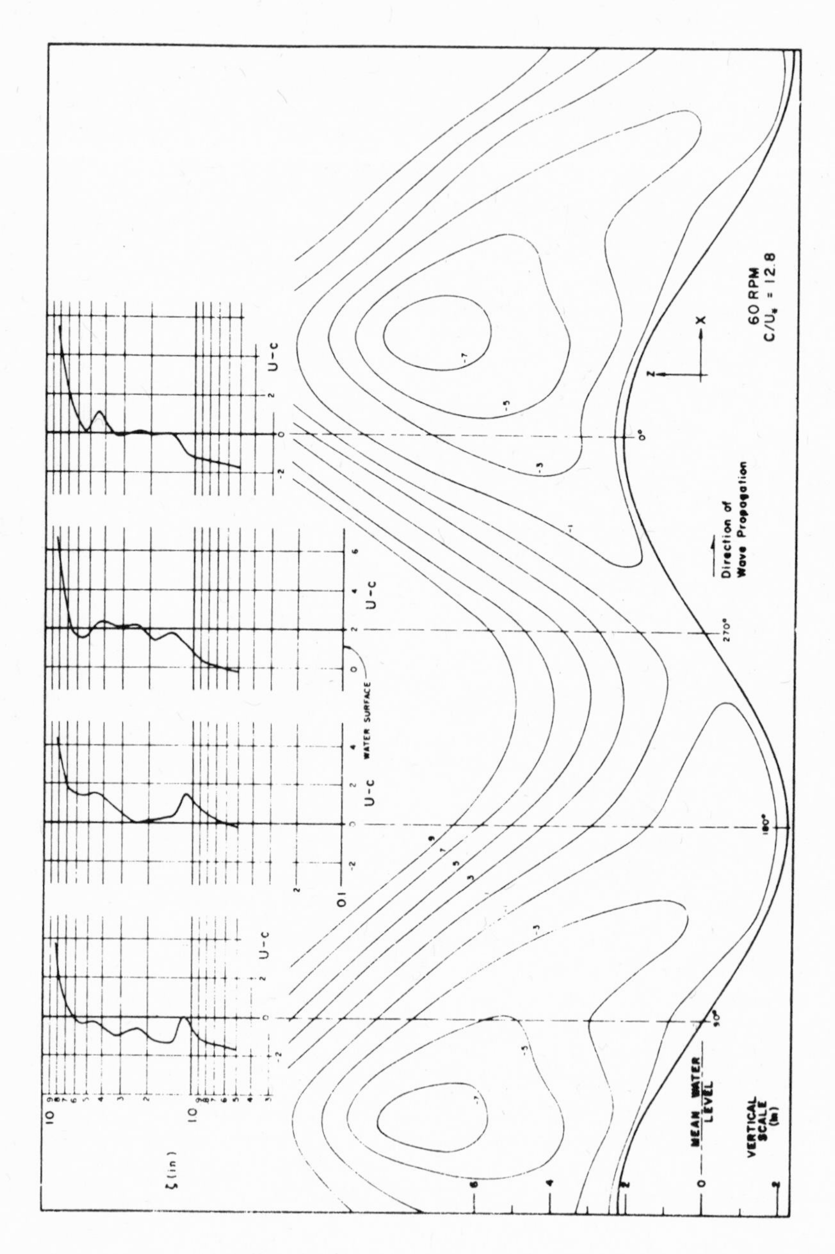

Figure 7. Wave-Induced Vortex Pattern in the Air Stream Over Waves.

$$M = \frac{b}{2} \rho_a g k a_o^2 \quad , \tag{25}$$

indicating that only the out-of-phase pressure coefficient accounts for momentum transfer. The latter depends on wind speed or shear velocity, wave phase speed and, to a lesser degree, on wave amplitude and turbulence level in the air stream.

The rate of energy transfer from air to water is given by

$$\frac{dE}{dt} = Mc = \frac{b}{2} \rho_a g \sigma a_o^2 \quad , \tag{26}$$

where E is wave energy per unit surface area,

$$E = \frac{1}{2} \rho_w g a_o^2 \quad , \tag{27}$$

and ρ_w is the water density. Equation 26 may be written as

$$\frac{dE}{dt} = b \frac{\rho_a}{\rho_w} \sigma E \quad , \tag{28}$$

where in this context b is customarily referred to as the wave growth coefficient and $\sigma = kc$.

The laboratory tests conducted by Shemdin (1969) with a wave follower included broad ranges of wave heights and wind speeds. The dependence of b on the ratio of wind shear velocity, U_*, to phase speed of wave, c, is shown in Fig. 8 for three different wave amplitudes. It is seen that b has weak dependence on amplitude and strong dependence on U_*/c. The solid line represents Miles' (1959) inviscid theory. The measured values of b have the same order of magnitude as Miles' theory in the range of $U_*/c \approx 0.15 - 0.4$. The theoretical transfer rates drop rapidly below this range and become negligible as wind speed approaches the value of wave speed.

Active wave generation in the field can occur when $U_*/c \approx 0.05 - 0.15$. The measurements by Dobson (1971), Elliot (1972), and Snyder (1974) are not in agreement on the magnitude of the wave growth coefficient. Dobson's measurements are one order of magnitude larger than Miles' theory while Snyder's measurements are lower but in general agreement with Miles' theory. Elliot's measurements fall between those of Dobson and Snyder.

The divergence between the various measurements prompted a joint experiment in the Bight of Abaco in 1974 and included instru-

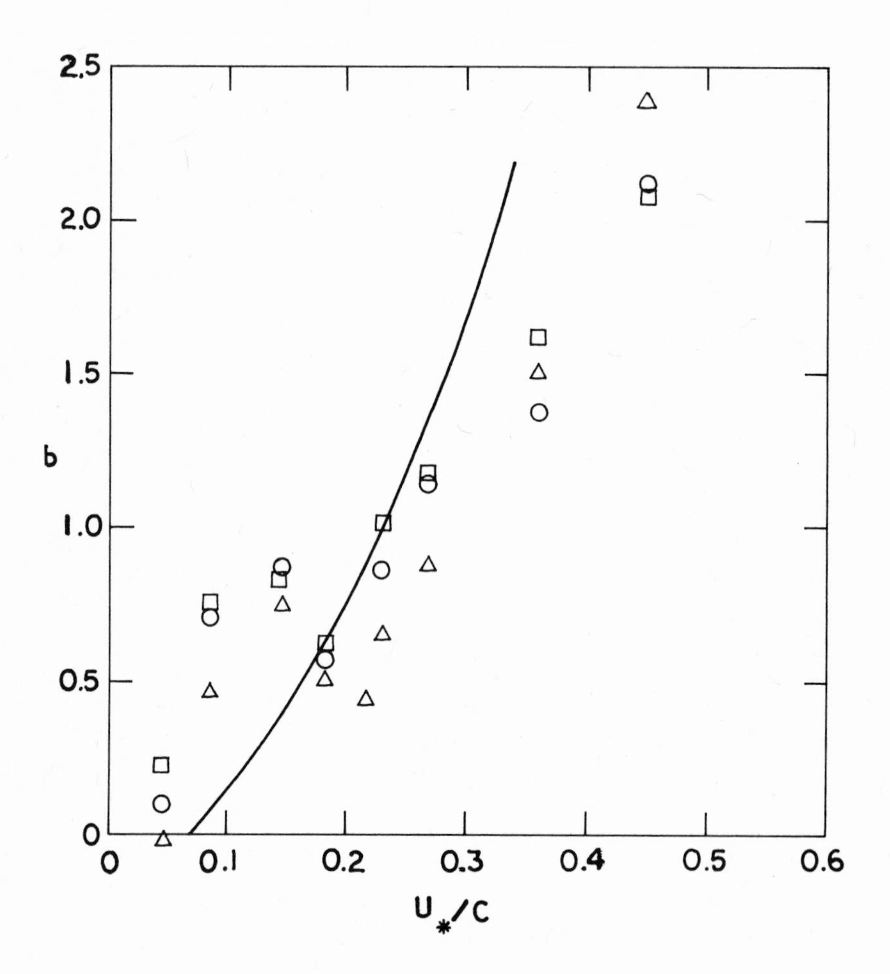

Figure 8. b vs. U_*/c for Different Wave Amplitudes and Comparison with Miles' (1959) Theory (Solid Line). △, a_o = 3.17 cm; ⊡, a_o = 5.38 cm; ⊙, a_o = 8.00 cm; Frequency = 0.78 Hz.

mentation systems provided by Dobson, Elliot and Snyder. The final results are not yet published but the early indication is that the growth coefficient is of the same order of magnitude as Miles' theory in disagreement with Dobson's (1971) results.

Progress on the theoretical scene was made by Townsend (1972) and Gent and Taylor (1976) who employed the turbulent energy equation to provide closure conditions for the momentum equations. For a uniform roughness height they found the growth coefficient to be consistent with Miles' earlier results. Gent and Taylor suggest that enhanced transfer rates can occur if the roughness height is allowed to vary along the wave profile.

Gent and Taylor (1976) compare available field measurements of the growth rate coefficient with their theoretical results (see Fig. 9). The theoretical ranges of transfer rates, shown as vertical lines, encompass transfers with and without roughness height modulation along the wave profile.

Snyder and Cox (1966) measured growth rates of waves by following waves in a boat traveling with the group velocity of the wave. The growth rates of waves on the forward face of the wave spectrum were found to be one order of magnitude greater than corresponding b values obtained from direct pressure measurements near the water surface. These results suggest that transfer by mechanisms other than direct atmospheric transfer occurs in this region of the wave spectrum.

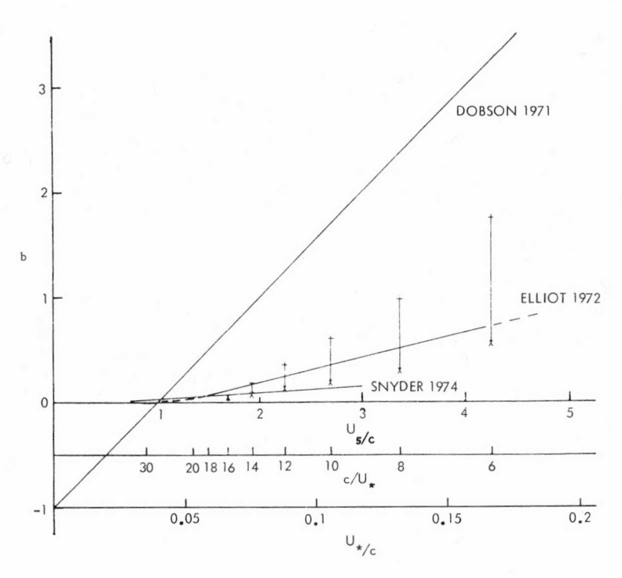

Figure 9. Comparison Between Theoretical and Experimental Wave Growth Coefficients, b, under Equivalent Ocean Conditions [after Gent and Taylor (1976)].

Hasselmann (1968), based on a weak interaction theory, suggested that nonlinear wave-wave interactions in a gravity wave spectrum can produce significant energy transfer between different spectral components in a wave field. Verification of this concept was found in experiments on wave growth reported by Hasselmann, et al. (1973). The suggested energy balance in a wave spectrum is shown in Fig. 10. This model suggests that rapid growth of waves is primarily due to energy transfer through wave-wave interactions, and to a minor extent to direct atmospheric transfer. The proposed balance also offers an explanation for the shift in peak frequency of the energy spectrum to lower frequencies. The net nonlinear transfer rate is maximum at frequencies below the energy spectral peak.

SUMMARY AND CONCLUSIONS

While details of mechanisms operating at the air-sea interface are not completely understood a number of significant investigations provide insight which can be used for engineering purposes. This is especially true for wind-induced drift where the presence of waves does not appear to induce a significant difference. The latter make it possible to model wind induced drift in the laboratory for complex lagoon geometries.

Momentum transfer to waves by direct interaction of waves with a turbulent shear flow above is the subject of active research. The growth rate of waves is determined by both atmospheric transfer from air to water and by nonlinear energy transfer across the wave spectrum.

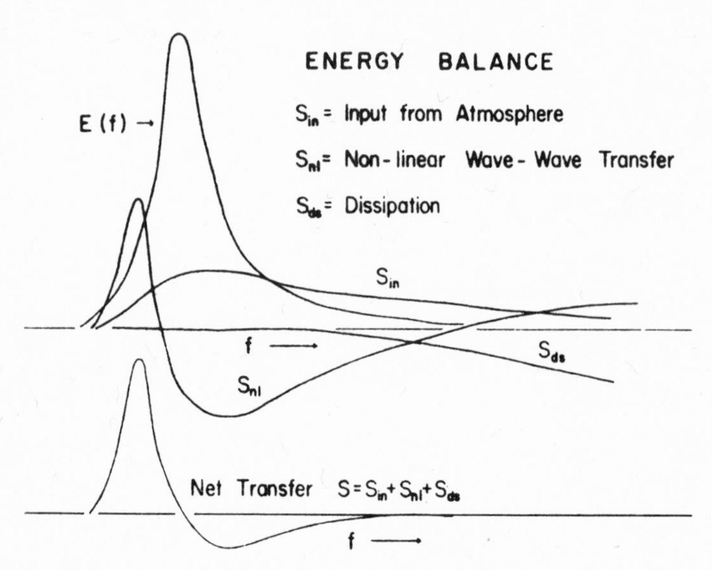

Figure 10. Energy Balance Model (after Hasselmann, et al. 1973).

NOTATION

a = in phase pressure coefficients
a_o = wave amplitude
b = out of phase pressure coefficients
c = wave phase speed
C_c = roof friction coefficient
C_G = group velocity of waves
C_z = drag coefficient of wind
g = gravity
κ = Von Karman constant
k = wave number
L = wave length
M = momentum transfer to wave
n_m = may correspond to laminar flow
n_p = may correspond to turbulent flow
η = wave amplitude
η_x = horizontal scales
η_z = vertical scales
σ = rms wave energy
p_a = aerodynamic static pressure
ρa = air density
ρ_w = water density
s = surface elevation above origin
τ_s = surface stress
U_{*a} = friction velocity
U_{*w} = water shear velocity
U_a = air velocity
$U_a(10)$ = wind speed at 10 m
$U_a(z)$ = wind speed at elevation z
$U_w(o)$ = surface drift
$U_w(-10)$ = current at a depth of 10 m
U_r = reference velocity at intake
ν_ω = kinematic viscosity of water

z = elevation

z_{oa} = roughness height above water surface

z_{ow} = roughness height below the surface

REFERENCES

Bye, J. A. T., "The Wave-Drift Current," J. Marine Res., 25, pp. 95-102, (1967).

De Leonibus, P. S., "Momentum Flux and Wave Spectra Observations from an Ocean Tower," J. Geophys. Res., vol. 76, pp. 6506-6527, (1971).

Dobson, F. W., "Measurements of Atmospheric Pressure on Wind-Generated Sea Waves," J. Fluid Mech., vol. 48, pp. 91-127, (1971).

Elder, F. C., D. L. Harris, and A. Taylor, "Some Evidence of Organized Flow Over Natural WAves," Boundary-Layer Meteorology, vol. 1, pp. 80-87, (1970).

Elliot, J. A., "Microscale Pressure Fluctuations Measured Within the Lower Atmospheric Boundary Layer," J. Fluid Mech., 54 (3), pp. 427-448, (1972).

Hasselmann, K., "Weak Interaction Theory of Ocean Waves," Basic Developments in Fluid Dynamics, M. Hoft, Ed., vol. 2, pp. 117-182, (1968).

Hasselmann, K., T. P. Barnett, E. Bouws, H. Carlson, D. E. Cartwright, K. Enke, J. A. Ewing, H. Gienapp, D. E. Hasselmann, P. Kruseman, A. Meerburg, P. Muller, D. J. Olbers, K. Richter, W. Swell, and H. Walden, "Measurements of Wind-Wave Growth and Swell Decay during the Joint North Sea Wave Project (JONSWAP)," Ergänzungsheft zur Deutschen Hydrographischen Zeitschrift Reihe A (8^0), Nr. 12, pp. 1-95, (1973).

Kenyon, K. E., "Stokes Drift for Random Gravity Waves," J. Geophys. Res., 74, pp. 6991-6994, (1969).

Keulegan, G., "Wind Tides in Small Closed Channels," NBS Res. Paper 2207, (1951).

Kitaigorodsky, C. A., and Y. A. Volkov, "On the Roughness Parameter of Sea Surface and the Calculation of Momentum Flux in the Near Water Layer of the Atmosphere," Bull (IZV) Acad. Sci. USSR, Atm. and Oceanic Physics, vol. 1, no. 9, pp. 973-988, (1965).

Lai, R. J., and O. H. Shemdin, "Laboratory Investigation of Air Turbulence Above Simple Water Waves," J. Geophys. Res., vol. 76, pp. 7334-7350, (1971).

Miles, J. W., "On the Generation of Surface Waves by Shear Flows," Part 2, J. Fluid Mech., 6, pp. 568-582, (1959).

Roll, H. U., Physics of the Marine Atmosphere, Academic Press, New York and London, pp. 131-140, (1965).

Ruggles, K. W., "The Wind Field in the First Ten Meters of the Atmosphere Above the Ocean," Department of Meteorology Tech. Report 69-1, M.I.T., Cambridge, Mass., (1969).

Shemdin, O. H., "Instantaneous Velocity and Pressure Measurements Above Propagating Waves," University of Florida, Coastal and Oceanographic Engineering Laboratory, Tech. Report No. 4, p. 105, (1969).

Shemdin, O. H. and A. J. Mehta, "Discussion on Paper by J. Wu entitled 'Anemometer Height in Froude Scaling of Wind Stress,' " J. of Waterways, Harbors and Coastal Engineering Div., ASCE, vol. 98, pp. 97-100, (1972).

Snyder, R. L. and C. S. Cox, "A Field Study of the Wind Generation of Ocean Waves," J. Marine Res., vol. 24, pp. 141-178, (1966).

Snyder, R. L., "A Field Study of the Atmospheric Pressure Field Above Surface Gravity Waves," J. Marine Res., vol. 32, p. 479, (1973).

Ursell, F., Chapter 9, "Wave Generation by Wind," Surveys in Mechanics, Cambridge Univ. Press, Edited by G. K. Batchelor and R. M. Davies, pp. 216-249, (1956).

Van Dorn, W. G., "Wind Stress on an Artificial Pond," J. Marine Res., vol. 12, pp. 249-276, (1953).

Wilson, B. W., "Note on Surface Wind Stress Over Water at Low and High Wind Speeds," J. Geophys. Res., vol. 65, pp. 3377-3389, (1960).

Wu, J., "Wind Stress and Surface Roughness at Air-Sea Interface," J. Geophys. Res., vol. 74, pp. 444-445, (1969).

Wu, J., "Anemometer Height in Froude Scaling of Wind Stress," J. of Waterways, Harbors and Coastal Engineering, vol. 97, pp. 131-137, (1971).

LIST OF CONTRIBUTORS

EDWARD M. BUCHAK, J. E. Edinger Associates, Inc., 37 West Avenue, Wayne, Pennsylvania 19087

JOSEPH DeALTERIS, Pandullo Quirk Associates, Gateway "80" Office Park, Wayne, New Jersey 07470

JOHN ERIC EDINGER, J. E. Edinger Associates, Inc., 37 West Avenue, Wayne, Pennsylvania 19087

M. I. EL-SABH, Section d'Océanographie, Université du Québec à Rimouski, 300 Avenue des Ursulines, Rimouski, Québec, Canada G5L 3A1

RONALD J. GIBBS, College of Marine Studies, University of Delaware, Lewes, Delaware 19958

WILLIAM N. GILL, Faculty of Engineering and Applied Sciences, State University of New York at Buffalo, Buffalo, New York 14214

JOHN GRIBIK, Basic Technology, Inc., 7125 Saltsburg Road, Pittsburgh, Pennsylvania 15235

E. R. HOLLEY, Department of Civil Engineering, University of Illinois at Urbana-Champaign, Urbana, Illinois 61801

H. P. HSIEH, Faculty of Engineering and Applied Sciences, State University of New York at Buffalo, Buffalo, New York 14214

C. P. HUANG, Civil Engineering Department, University of Delaware, Newark, Delaware 19711

ROBERT T. KEEGAN, Pandullo Quirk Associates, Gateway "80" Office Park, Wayne, New Jersey 07470

GI YONG LEE, Faculty of Engineering and Applied Sciences, State University of New York at Buffalo, Buffalo, New York 14214

T. S. MURTY, Marine Environmental Data Service, Ocean and Aquatic Sciences, Department of Fisheries and Environment, 580 Booth Street, Ottawa, Ontario, Canada K1A 0H3

FLETCHER OSTERLE, Department of Mechanical Engineering, Carnegie-Mellon University, Pittsburgh, Pennsylvania 15213

RALPH R. RUMER, JR., Civil Engineering Department, University of Delaware, Newark, Delaware 19711

RICHARD P. SHAW, Department of Engineering Science, Aerospace Engineering, and Nuclear Engineering, State University of New York at Buffalo, Buffalo, New York 14214

OMAR H. SHEMDIN, Jet Propulsion Laboratory, California Institute of Technology, 4800 Oak Grove Drive, Pasadena, California 91103
WILLIAM J. SNODGRASS, Department of Chemical Engineering, McMaster University, Hamilton, Ontario, Canada L8S 4L7

INDEX